MW00981926

Also by Tom Henry

The Good Company:
An Affectionate History of the Union Steamships

Westcoasters:
Boats That Built BC

Small City in a Big Valley:
The Story of Duncan

Inside Fighter:
Dave Brown's Remarkable Stories of Canadian Boxing

The Ideal Dog

Dogless in Metchosin

Paul Bunyan on the West Coast

FOLLOWING the BOULDER TRAIN

Travels with Prospectors and Rock Doctors

Tom Henry

HARBOUR PUBLISHING

Harbour Publishing Co. Ltd.
P.O. Box 219
Madeira Park, BC V0N 2H0
www.harbourpublishing.com

Cover photo by Dan Shugar
Printed and bound in Canada

Harbour Publishing acknowledges financial support from the Government of Canada through the Book Publishing Industry Development Program and the Canada Council for the Arts, and from the Province of British Columbia through the BC Arts Council and the Book Publishing Tax Credit.

Library and Archives Canada Cataloguing in Publication

Henry, Tom, 1961–

Following the boulder train : travels with prospectors and rock doctors / Tom Henry.

Includes index.
ISBN 1-55017-377-4

1. Miners—British Columbia—Biography. 2. Mines and mining—British Columbia—History. I. Title.

TN140.H445 2006 622.092'2711 C2006-903336-6

Contents

YUKON
TERRITORIES
NORTHWEST
TERRITORIES
Atlin
Atlin L
Cassiar
Juneau
Thorn Property
ALASKA BORDER
Dease Lake
BRITISH
COLUMBIA
Eskay Creek
Hyder
Stewart
Ketchikan
Takla Landing
Kispiox
Hazelton
Prince Rupert
Terrace
Smithers
Houston
Fort St James
Kitimat
Burns Lake
ALBERTA
Vanderhoof
Prince George
Quesnel
Bella Coola
Williams Lake
PACIFIC
OCEAN
Clinton
Kamloops
Campbell River
Powell
River
VANCOUVER
ISLAND
Boston Bar
Merritt
Port Alberni
Squamish
Kelowna
Princeton
Nanaimo
Hope
Oliver
Victoria
WASHINGTON
0 km
200 km
Approx. scale

Introduction

The price of a troy ounce of gold the year I began writing this book loitered at $326. It had not moved significantly in months. In times of a buoyant economy and ascendant gold values, $326 troy would be high enough to stimulate exploration, to prompt mining companies and investors to focus dollars into the risky endeavour of finding gold and other minerals.

But something had interfered with tradition: that in a time of insecurity, the dollar falls and gold rises. Was it investor mistrust after Bre-X, the Borneo gold scam in which investors lost billions of dollars? Post-September-11 anxiety?

Some people said that investment money was slurped up by the dot-com failure. In a normal world, at times of collective jitters, the dollar went down and gold went up. The mechanism of a teeter-totter.

But times were not normal. In Vancouver, the world centre of mineral exploration, empty office chairs were tucked under blank computer screens, and the few geologists who could find work schemed on how to make projects last rather than how to fit them all in. Many geologists, tired of cleaning house while their wives pulled in cheques teaching sociology, took up carpentry or learned website design. And in the Interior, in towns like Dease Lake and Ross River, the locals who relied on exploration work stared at the quiet phones and wondered what they would do for work that year. It was, I was told again and again, a strange year, and not at all the sort on which to base a book about people who look for valuable rocks.

Ever since I was a boy I wanted to work in the elemental industries of the province: farming, logging, shipping, mining. I grew up on a grain farm in northern BC and I live on a small farm on southern Vancouver Island; during my twenties and thirties I worked in a logging camp and on tugs. What I had never done, however, was work in mining or exploration. Writing this book was a chance to live out one of my long-held ambitions.

It was also a chance to explore an industry that has backboned BC history. The single most iconic photo of BC—of a horse-drawn wagon wending the Fraser River trail during the gold rush—speaks to the seminal role minerals and mineral exploration played in the province's development. British Columbia came into existence as a direct result of the discovery of gold on the Fraser River and in the Cariboo in the 1850s and 1860s: the hoards of American gold seekers that arrived as a result forced the colonial government to extend British sovereignty over lands previously controlled by the Hudson's Bay Company. Formed to implement mining legislation, the new government went on to establish the laws on which the province is based. All because gold was discovered.

Not only did mining trigger the formation of BC, but it has also been one of its top three industries for 150 years. It was the founding instrument of countless communities and remains, in dollar value, the third-largest industry in BC after forestry and tourism (with the inclusion of oil exploration, it is the number-one industry). It annually generates $6.2 billion and employs about six thousand people. Shadowing prospectors and writing this book would offer me a chance to explore, at least in some way, the role mineral exploration has had in BC's development.

I was interested, too, in the way geologists think. The ultimate treasure hunters, they combine an understanding of physics and chemistry with Boy Scout lore to search for minerals. Your typical exploration geologist is adept with a magnetometer, conversant in the ways of surly grizzly bears and antsy investors, can predict a change in weather by the force of a katabatic wind, knows how to temper a rusty skillet, can stride up and down mountains all day and, with only pasta, beans and a little oil, can whip up a feast on a camp stove in a howling southeaster. Exploration geologists think in four dimensions and speak a language that is both unique and at times as beautiful as what it expresses. They think of millennia as I do minutes and express massive tectonic contortions as if the gliding continents were bumper cars. Describing the formation of BC to me, one geologist used the words "smashed," "splat," "flipped" and "smushed."

In the story of mining, nothing is as pivotal as the role of the finders, the people who use their wits and dogged determination to suss out where, beneath the dirt, gravel and bush that cover them, valuable minerals lay hidden. The most glorious positions in the pantheon of mining heroes are occupied by these discoverers, these prospectors and rock choppers. The name of the "Coal Tyee," Snuneymuxw chief Ki-et-sa-kun, lives on as the man who showed Hudson's Bay Company where the great coal beds were at Nanaimo. If the average British Columbian can remember the name of one miner connected with the Cariboo Gold Rush, it is probably that of Billy Barker, the man who discovered the rich diggings on Williams Creek near Barkerville. Likewise, if there are any names most people can dig out of the historical sediments obscuring the Klondike Gold Rush, they are probably those of George Carmack and Skookum Jim Mason, the prospectors who made the discovery strike on Bonanza Creek.

If the mining industry has a cultural icon to put up beside those of the cowboy, homesteader and lumberjack, it is that of the prospector, the grizzled old bushman with his backpack and gold pan on his back, rock hammer in his hand, spending his days wandering the backwoods hacking at rocks, talking to bears and dreaming of the big strike. This is the romantic figure that inspired me. What has time wrought with that solitary soul? Who carries on his (or her) crucial work, without which we could not have our titanium glass frames, chrome-plated bathroom faucets or aluminum jumbo jets? I knew the heirs of this great tradition of mineral-hunting must be out there somewhere, but how did they carry on their work today? Did they still maintain their treasured independence, or were they all organization men (or women)? Did they still wheedle their grubstakes and heft their Trapper Nelsons, or was it all electronics and helicopters today? These were some of the questions that drew me forward.

When I began doing research for *Following the Boulder Train*, I was told that I had missed the boom years—the late 1980s when Eskay Creek was being opened up and the likes of Murray Pezim were fastballing money around. Hell, an enterprising individual could make a small fortune supplying staking posts. Mineral exploration firms hired from the back of the dingiest bars—so what if the old prospector spent the first few days ape-ing around in tree branches, claiming to see Colonel Sanders dispensing three-piece chicken dinners? In a few days he'd square away.

Or I was told I should have seen the boom years of the 1970s, when helicopters swarmed over the interior valleys like blackflies, and budgets were so generous that the camp boss might airlift in a loaf of fresh bread, even though the chopper billed out at hundreds of dollars per hour. Twenty-five dollars a toast-slice? Who cared!

The epitome of that great era may have been an unofficial exploration expo held in 1975 at the Lapic Campground on the way into Ross River. No less than thirty-two helicopters flew in, each jammed with crews, for a big celebration. Events included gold-panning, claim-staking and plucking hundred-dollar bills from the treetops while hanging from a helicopter. I can't be sure, but apparently one buck pilot did a barrel roll. Someone brought in a former *Playboy* playmate as camp cook. It was a kinetic, crazy, delusional time.

I was delighted not to be part of that era. I've tried before to write of bacchanalian high living and always failed. Excess and energy and pioneer optimism are better suited to the atten-tion span of the camera than the long steady stare of narrative. People stick out their chests in good times, rooster around and hot-roast the truth. I prefer things at a lighter, more reflective temperature. Even in the time that I was working on this proj-ect the price of gold went up and some of that useful humility I had seen vanished. Talking with explorationists and prospectors then and now was as different as talking to a man on a plane that's lost an engine and a man who's flying first-class with the movie on.

So the timing for the book was good.

What remained was where and how. Though exploration geology occupies a small portion of the discipline of geology, it is an immense field, incorporating the disciplines of chemistry and physics and the techniques of financing, politics and economics, to name just a few. It is an enormous and challenging discipline and to think that I, a general-interest writer, would be able to do it justice in any sort of proportional way would be delusional. Yet there seemed to be a solution. Northern British Columbia literally and figuratively is representative of the exploration industry. Along with the Andes in Chile, and the mountains of Borneo, it remains a most promising mineral-rich area. The challenges of exploration here in some small ways mimic the challenges of early exploration, even in an age of satellite phones and helicopters. So that to wade through the devil's club at the Thorn property, as I did, was to at least sense the difficulty and elation of eighteenth- and nineteenth-century explorers.

Finally, that left the question: how to present such a story? I dismissed the idea of writing a history right away. There are many histories of early mining, especially as they relate to gold rushes. Creating a belated addition to that canon was not my ambition. Furthermore, mining is a live-in-the-moment industry, in part because the chances of success are so low by ordinary standards. A figure often tossed out is "one in four lifetimes"—for every four geologists, one will find a significant ore body. Expressed cynically, that means three geologists will search all their lives and find nothing. If geologists thought about history and probability they'd leap from cliffs.

Nor was an overview of the exploration industry in order, because it was people I was after and not business. If some of the locomotive events of business crept in, so be it.

In the end, the form for this book was shaped by the industry itself. One of the most basic prospecting skills is the ability to follow what is called a boulder train. A boulder train is the leavings of a glacier that has scraped off material, moved it and then deposited it as the glacier melts. It is an entry-level skill that can be forever improved, like tracking an animal. It is nose-to-the-ground, trust-your-instincts stuff. I decided I too would follow something big, not knowing where it would take me or what I would find.

A Trip North

Through July and three weeks of August I awaited the call to head north. Then, when it finally came, what did I do? Balked. Tried to jam out. The garden, I suddenly knew, needed tending. Bindweed and thistle and dock mobbed the vegetables. Tansy ragwort had budded in the compost. And the sheep seemed to require immediate footwork. Number 20 limped and Number 11 was lying down more than others in the flock. Ingrown hooves? I really needed a week to tend my small farm. Furthermore, my daughter was about to enter a new school and my presence for after-school ice creams seemed important. It was a lousy time to leave home.

A dozen and more excuses rolled through my head even as I answered the phone. It was Henry Awmack, a geologist and principal in two Vancouver mineral exploration companies. Months earlier, Awmack and I had met in his seventh-floor corner Vancouver office to discuss the possibility of writing a book on geologists and prospectors. Well over six feet tall, and with a bristling red beard, Awmack had an abrupt countenance completely at odds with his character. He looked snappy and surly, yet he was soft-spoken, hesitant and genial. He padded around his rock-strewn office in socks.

After some discussion, he agreed to let me join several projects in northern BC. It was a generous offer because mineral exploration firms are both proprietary about their projects and, having seen the industry repeatedly labelled as malicious exploiters in the press, are wary of publicity.

The mineral exploration industry—the people who actually seek ores—have long been the unsexy sibling to the flimflammery of mining promoters. Awmack hoped for an even break. "Have a look at what we do. Talk to some people," I recall him saying. I had no agreement with Awmack as to how and what I would write, but I had said I would, as much as possible, try to view any controversial topics from a number of sides.

From earlier conversations with Awmack, I knew I had a chance to witness what may well be a vanishing industry—and with it a vanishing type of work. Though it has since recovered, in 2001 the exploration industry was mired in what is regarded as one of the worst bear markets in memory. Exploration budgets had shrunk dramatically since the heady early 1990s—from a high of $230 million per year to a paltry $32 million. Sure, the industry was used to cyclical gyrations, but there were about the malaise some worrisome signs of permanence. Investment money, always necessary in large quantities for mineral exploration, had moved to the technology sector. The exploration industry worldwide was still bruised by the Bre-X scandal. Chile, Argentina and Peru were developing mines and are, in many ways, where BC and Canada were thirty years ago—on the leading edge of a mining boom. And, with world population predicted to flatten out within a generation, the need for evermore mines seemed finite.

The collective effect of these large-scale factors was to make the unique and small group of exploration geologists and prospectors in BC feel as if they were perhaps the last generation of their kind. Universities were producing geologists attuned to the burgeoning field of "environmental impact" studies, and those who were interested in precious metals sought out diamonds, which are a different game, played in a different arena. In the Vancouver offices of mineral exploration companies it wasn't unusual for the youngest geologist to be forty or even fifty years old. Unless something changed quickly, the sentiment was at the time, there would be no "next generation" of mines because there would be no next generation of geologists to find them.

Our plan had been for me to go north in June, but investment money was tight and the season was coming to a close. If exploration was going to happen, Awmack reminded me on the phone, things had to move quickly. Snow could fly in the mountains of the Iskut Valley, where I was bound, at anytime. "Bring warm clothes," Awmack clipped. "Be prepared for snow. Rain. Whatever." He was hurried. He told me to meet a helicopter at kilometre 45.5 on the Eskay Creek haul road off the Stewart–Cassiar Highway, and gave me a date and time. He hung up before I could tell him about my needy sheep, my parental duties.

I had already talked to enough geologists and prospectors to know that pre-trip trepidation is a common affliction, just as stomach butterflies are for hockey players before they step on the ice. However much geologists enjoy the bush, there is a nasty time just before they leave home when, if a brother-in-law were to offer them shares in a profitable shoe store, they'd ditch the whole rock business in favour of hawking Birkenstocks. Teaching seems like a better profession. "I hate the days before I go into a camp," geologists had said to me again and again.

Why did they hate the idea of going so much, and why did they keep returning? Of many reasons offered, I concluded that the negatives—particularly the duration of work shifts—are completely balanced out by the positives—the appeal of the lifestyle. Because of a provision in provincial labour codes, salaried geologists and field crews can go into the field indefinitely and under circumstances that would never be permitted in hourly-pay jobs. There are no provisions for overtime. Geologists in the middle of their careers told me about entire summers in the bush, when respite was a quickie chopper ride from a tent camp in nowhere to the bar in Ross River, Yukon, for a couple hours of heavy drinking. Even though the era of four-month shifts is gone, it is still common for geological exploration field crews to spend six and eight weeks away from home. In our "right now!" world, that is too long.

Strangely, geologists are often the first to defend the status quo. They say that the expense of getting crews in and out of remote locations precludes regular shift changes. But miners and loggers who work in equally remote areas regularly commute. I knew that miners at the Eskay Creek gold mine in northern BC, for example, work two weeks on and two weeks off. Many fly home to New Brunswick on their days off. Meanwhile, up on a wind-stripped ridge above the mine, a geologist with a PhD is entering in his journal, "Week four. No shower or bath . . ." Under those kind of conditions it is difficult for geologists to look forward to leaving home.

From those same geologists I had also heard about the appeal of their work. "Once I get in the bush I love it," said Frank

Gish, a geologist based on Bowen Island. "You remember why you got into the business in the first place. It's what it is all about." Again and again geologists described to me the twin halves of very distinct working lives: at home, when they—like the rest of us—commute to carpeted offices, get too many emails and joke about the hockey pool results; and the life in the bush, where bears and hundred-kilometre views are everyday occurrences, home is a tent at three thousand feet beside a glistening tarn and working hours are spent intuiting the earth's secrets in what amounts to a massive treasure hunt. It was the in-between part that they hated. Though my trip was just beginning, I could understand why.

Monday. A fretful launch to the journey. I woke at 2 a.m., 3 a.m., 4 a.m., to unload subconscious worries. Every anxiety seemed to include a gysering waterline. My greatest fear is being unprepared. The night before I loaded my Ford Ranger with a mule packer's fussiness. I burden trips with so many expectations that to arrive somewhere and find I haven't packed even a small item is to endanger the whole enterprise. As if I can't buy a penknife in Prince George? A hat in Hazelton? The answer would be to put everything in the truck the night before leaving, and then walk out buck-naked in the morning and dress *inside* the vehicle. That way I wouldn't forget a thing.

The morning was warm and the sky clear but for a few fleecy-looking clouds surfing overhead on a late summer westerly. I waved goodbye to the dog and pulled onto the road. I had plenty of time to catch the first ferry. Silly fishermen were headed the other way, their SUVs sucking hard to pull trailered Bayliners. I felt good. Yet within five minutes I realized I had a problem: the Ford's heater had croaked. It blew only frigid air. Faced with a humble return home, or driving with an extra sweater, I opted for the latter. I hoped that I had had my obligatory mechanical problem. And that temperatures in the days ahead wouldn't dip below freezing.

Vancouver Island is a lousy home for true adventurers. You leave home for the mainland enthusiastic about the day ahead; then, forty-five minutes later, you find yourself in the ferry lineup, slumped behind the steering wheel in dismal ennui. You board the ferry, wait again, get off and wade through several hours of Lower Mainland traffic. Five hours after you leave your Vancouver Island home you are maybe thirty miles into the trip. In many cases, the spume from your hometown pulp mill is visible in the rearview mirror. Not a wonder the island has a reputation as a refuge of the timid. Any serious adventurer would have abandoned it years ago. I wonder how many marriages have been saved by the Georgia Strait crossing? A husband leaves his wife in a huff and sets out for a blousy old flame that runs a video store in Moose Jaw. He misses the 7 p.m. sailing, thoughtfulness sets in, and two hours later he's back in his Ladysmith rancher making up.

The Americans on the ferry had big, undefined crotches, like Dr. Seuss characters. As docile as cattle, they knotted around the closed entrance of the cafeteria until one—an elderly woman in an oversized Gore-Tex jacket—spotted a gap in the retaining ropes and slipped through. Six of the dozen nervously followed until they belatedly realized—surprise!—the signs were right; the cafeteria wasn't open. They stood, helpless-like and unsure of how to reunite with the others, until a surly manager came along and opened the ropes for them. I expected one of them to moo in excitement of reunion.

Midway across the strait, I could sense a generous, big-hearted feeling coming on. Never mind the wonky truck heater, never mind the dinks, I wanted a real crisis on this trip. Hoped for it, even. A flat tire in Alexandria, a bottle of scotch with the gout-hobbled, one-eyed prospector who stops to help me . . . who knew, maybe I'd end up with 50 percent shares in an ore-rich claim called "Teenage Crush"!

Of the pleasures of travelling fast on the Fraser Valley freeway I'll leave to someone else to describe. Perhaps it has something to do with horsepower and good CDs. My aging Ford Ranger had neither. I drove through Chilliwack angrily stabbing at the radio buttons in search of a voice other than Shania's. Farmland extended to the down-sloping mountains and met with what looked to be a one-pace transition. You could be hoeing pumpkins on Fraser Valley bottomland and bang your elbow on the adjacent rock face. As the valley narrowed I passed a billboard whose message was to my liking. With the deft addition of a spray-painted "r," vandals had transformed "Praise God" to "Praise Gord."

Somewhere out of Hope the radio signal expired with a sound like the last transmission from a foundering ship. The traffic thinned. Ahead of me a motorcycle and sidecar was having trouble with hills and curves (that's all the Fraser Canyon highway is) and could do no more than the speed limit. Every time they turned right the passenger and driver studied the sidecar for signs that it was about to tip. The highway was built in an era when engineers made concessions to geology, not like the Coquihalla, where the four lanes bull through. The hills in the Fraser Canyon have runaway lanes, some marked with the calligraphy of locked-up tires, a reminder that not all trucker stories are bullshit.

As I drove, it occurred to me too that at one time or another all the great prospectors had come this way: Billy Barker, for whom Barkerville is named (by accounts, a real asshole); Bill Dietz, who discovered gold at Williams Creek in the early 1860s; H.H. "Spud" Huestis, who bulled the great Highland Valley copper deposit into production in the 1950s. The road into the province's Interior was redolent with their ambitions.

An Iskut-bound team of hard-driving geologists in a big-block Ford F350 can make it from Vancouver to Smithers in fourteen hours. They'll sleep at the Fireweed Hotel on Main Street in Smithers (where sacks of rock samples no longer puzzle the staff) then rise before dawn to be at the Eskay Creek turnoff by early afternoon. If the weather co-operates and the contract helicopter is on time, they can fly in their gear, set up camp and be perched on plastic chairs and eating veal cutlets and mashed potatoes by evening of the second day.

With five days to do the same trip, I decided to use my time in loose study. Never fond of long hours of driving, I would pull over, eat and drink and stare at a local feature, then relate it to something in a reference book. My ambition was to familiarize myself with the ABCs of the province's geology, so if a geologist mentioned the Intermontane Belt, or the Cache Creek Terrane, I'd know what they were talking about.

To keep supplies to a manageable minimum, I had winnowed masses of geological reports and books to three items: an oversimple but highly portable series of pamphlets on the roadside geology of British Columbia; an interesting if idiosyncratic volume on the province's geology entitled *Where Terranes Collide* by retired Geological Survey of Canada geologist Chris Yorath; and the American Geological Institute's *Dictionary of*

Geological Terms, which I used to not only decipher the often incomprehensible language of geology I found in reports, but to paw through when I had nothing else to do (favourite word to date: "vug": surely one of the few words descended from Cornish, and meaning, according to the dictionary, "a small cavity in a vein or in rock . . .").

A former claim-staker-cum-farmer had loaned me the roadside guide. It was four folders, each for an east/west swath of the province. Clearly, the challenge for its creators was to present the map in such a way that was comprehensible to people who move (and often think) about the province in east/west coordinates, and yet not do too great an injustice to the geology, which is predominantly north-south.

I pulled over at a stop in the Fraser Canyon, spread a lunch and studied the books. The mass of mountains between the west coast of Vancouver Island and the Alberta foothills is part of what geologists call the Cordillera. *Cordillera* means "rope" in Spanish and refers to the mountains that extend along the western edge of North and South America. Like many words in geology, it is also deployed as a firewall between those who understand the discipline and those who don't. To march into a map store and ask for a map of the "Kordiller-eh?" as I once did, is to invite derision. "Cordillera" rhymes with "high sierra."

The most dramatic physical expression of the Cordillera is the mountains—the Coast Mountains, Cassiar–Columbia, the Rockies. These mountains rise from what geologists have identified as unique belts. From west to east they are the Insular Belt, the Coastal Belt, the Intermontane Belt, the Omineca Belt and the Foreland. But these are only expressions of still larger subterranean assemblages that geologists generalize as terranes, and which they represent on maps in long, narrow north-south trending bands that look approximately like a hand, palm-down, fingers together, on a table. Terranes differ from terrains: a terrain is what's on the surface, or topography; a terrane is a three-dimensional chunk of earth. And it is at the level of terranes that geologists get excited, because it is at the level of terranes that the province was assembled. To geologists, terranes are chapters in the story of BC.

According to geologists like Yorath, the story of the creation of BC is both quick and violent. It begins about two hundred million years ago, which is like last week in earth's five-billion-year history. In geological time, the tuna salad in a geologist's fridge wouldn't have gone off in two hundred million years. An ancient continent, Pangaea, had just broken up and its component pieces were heading off in all directions around the globe, propelled by the conveyor-belt of plate tectonics. What is to become the core of North America rifts westwards. It is old granite. The site of the present-day Saddledome in Calgary is waterfront on an ancient western ocean. Far to the west are a number of terranes called island arcs; they look similar to the Aleutian Islands. An island arc is formed when a section of sea floor dives under another plate. At some point under the plate, the subducted material is superheated, and gives off magmas that rise in balloon-shaped bubbles. If they break free they form volcanoes, like Mount Baker. If they don't break free they form batholiths, huge underground masses of rock. The Coast Mountains, which had just sent a shard of shade over my picnic table, are the biggest batholith in the world.

I finished my drink and packed the remains of lunch in the cooler. Traffic was sparse, only transports and a few motorhomes. Chris Yorath's descriptions of continental movement, I decided, could be nicely illustrated using traffic. The continental craton was a westbound logging truck; the eastbound terranes were Winnebagos with fat women clutching chihuahuas. According to the modern understanding of how the province was assembled, the first Winnebago hit the continent broadside. The Foreland belt was emplaced. Behind the accident, somewhere out on the Pacific, drivers panicked and swerved. One, two, three motorhomes rubbed together, came drifting into the continent as one messy terrane. The Intermontane Belt was emplaced.

Contact of such dimensions is not without consequences. Eastward blocks of terrane were driven over other terranes, sometimes at startling speeds. In the conflagration of Winnebagos, bumper melded with bumper. The great pressures produced metamorphic rocks. That's the Omineca Belt. Sediments scraped from the plain were squeezed and slid around as sedimentary rocks. This was a massive pileup, one that would have led the six o'clock news. I bucked up the canyon in my little truck. Even though I'd driven the route dozens of times before, I felt like it was all new. And in a way it was.

I purchased thirty dollars of ninety-four-octane gas and a Coke in Boston Bar. Back on the highway, the heavy, late-summer heat buffeted my elbow. The road was enveloped in black and brown rocks. The traffic was light, just a few trucks and regional Datsuns with three cylinders—good for a trip to the store and back. Inexplicably (marvellously?), three teenage girls appeared on the roadside, walking from nowhere to anywhere. This rock business, I decided, could get oppressive. The damn things were everywhere. I felt like I was travelling in a foreign land and I couldn't understand a word of the language. I stopped at the roadside and consulted the dictionary. What the hell is a rock, anyway? "An aggregate of one or more minerals," it said.

Minerals are discrete and therefore easier to identify. Rocks do not have rigorous compositions.

Through Clinton, and up the great hill to the east. My little truck was labouring in third gear, thinking about second. I was going through what geologists call a physiographic shift. Coastal Belt to the west, Intermontane Belt to the east. I was heading into the plateau country of snake fences, logging trucks laden with pecker poles, towns where the lead item on the front page is some sort of half-separatist rant about Victoria. One of the more flighty theories about geology suggests that people are influenced by the type of terrane they live on. Hence staid and solid cultures tend to live on cratons—huge ancient chunks of ancestral continents, of which the Canadian Shield is one—while creative versatile cultures live on new terranes—like California and much of BC. According to what I'd read in Yorath the theory seemed to have some validity. Central BC, which is given to political outbursts, is largely a volcanic eruption.

It was 6 p.m. when I arrived in Quesnel, a tidy little town of wood and cinder-block buildings set in a saddle around the junction of the Quesnel and Fraser Rivers. The town doesn't fawn for tourist dollars like Chemainus, nor does it aspire to something it isn't, like cosmopolitan-wannabe Burnaby. In many ways it is the most British Columbian of towns. The smell of pine from the sawmills was everywhere. Petunia-jammed gardens lined the sidewalks. The girl behind the counter at Shoppers Drug Mart had biceps the size of oranges and a horse pendant around her neck. The kid at Petro-Can sported an off-green shiner.

During the 1930s, when other cash-poor towns in BC were shunting the homeless onto the next town, the good people of Quesnel quietly welcomed them. They came because of the river, and the modicum of gold it would yield to a man who would work it. Most Depression-era gold panners averaged a few dollars a day for their hard work, but it was enough for a man to lay in provisions for freeze-up, to buy a can of McDonald tobacco.

I booked into the Cariboo Hotel, an enormous barn of a structure with out-of-kilter doors and windows and a drive-through cold beer store. I tried to make conversation with the woman behind the counter as she juggled beer sales and room bookings. I said, "There's so much traffic I thought there must be something special going on, but then I remembered it was the weekend." She said, "It's Quesnel. It's always like this."

Later, I walked across a wooden bridge and had dinner on the porch of a pub overlooking the town. A group of loud, big-shouldered men and their big-haired wives downed blue drinks. A hot rod went by then another and another. All glowing chassis and frame, they looked like big-titted anorexics. They were members of the Prospector's Car Club. Even though it was dusk, their souped-up buggies glowed red and yellow and green.

Day two. By the time I turned west onto the highway to Smithers the next morning, I had seen, if not driven over, the major terranes in BC. The highway junction is in Prince George. Prince George sits on an eastward-projecting nub of the Intermontane Belt that protrudes into the Omineca Belt. In my rearview mirror, rising like teeth on the horizon was the Rocky Mountains, the expression of the Foreland Belt. After that it was prairie on craton, the ancient edge of the sea.

To finish the story, then: When the superterrane smacked into the continental shelf, it did not go under it. Rather, large pieces of the Quesnellia and Slide Mountain terranes peeled off the oceanic plate and overrode the continental margin. Some slices were twenty-five kilometres thick. The impact folded the rocks of the Superterrane and formed the Columbia, Omineca and Cassiar Mountains. Rocks in some of these mountains are metamorphic. Impact bulldozed the layers of sedimentary rock that covered the continental core eastward. The image geologists often use is of a carpet being pushed. The folds are mountains. When the pressure became too great, limestone layers broke, becoming, in the lingo, thrust faults. Blocks of rock above a break are called a thrust sheet. Those sheets—toothy, white, were what I was looking at in my rearview mirror—the Rocky Mountains.

I arrived in Smithers shaky, cranky, redoubled in my commitment not to become a truck driver. Even though it was Labour Day weekend and the town was full of farmers and ranchers attending the Bulkley Valley Fall Fair, I was able to rent a room for fifty-nine dollars. Gouging did not seem part of the town's lexicon. I had in my pocket a folded piece of foolscap with the names of people I might talk to in Smithers: Tom Bell, Lorne Warren, Pat Suratt. Good, what-can-I-do-for-you names. I started dialing. Everyone was out or, incredibly, his father-in-law had just died. It was a bad season for fathers-in-law.

The street outside the hotel was being resurfaced, and the hot asphalt smelled of horse urine leaking from passing trailers. It was the last day of the fair. It was the biggest event for hundreds of kilometres—whopping 4x4s drawing four- and six-horse trailers. Some of the distant entrants were leaving, wives in the passenger seats cupping their hands to the wind to jib the breeze, kids in the crew seats already travel-tired. I showered and drove back to the fairgrounds in time to watch the horse-pull competition. Afterwards I wandered into the displays of fruit and flowers. Clearly, the domestic arts not only endure but thrive. There was, however, no celebration of mining.

Tom Bell was nursing a nasty hangover. A big-boned man, with fighter's knuckles and receding blonde hair, he attacked his sausage and eggs as if they might offset something going on in his stomach. I had heard a story about the time Bell met a mining promoter who had not paid him for prospecting work. Bell met the promoter at the Jolly Taxpayer Pub in Vancouver, and clutched the edge of the table with his beefy hands in such a manner as to suggest that, should the promoter not pay up soon, expensive dental work would be forthcoming. Bell was handed the cheque the next day. Looking at him across from me in the Hudson Bay Lodge, I could see why. His shirt stretched to span his broad shoulders. His attitude, though friendly, seemed to layer a quick temper. Both Bell and his partner Diane, who was at his side, were deeply tanned. They had been to a beer garden yesterday, then a dance. They seemed blissfully exhausted.

Bell was just back from the Yukon, where he and a geologist had been on a reconnaissance of five areas. They drove a 4x4 into the bush as far as they could go, then left the truck on a sandbar and radioed for a chopper to take them farther. While the geologist mapped an area, Bell looked for samples. When he works, he says, he's on all fours, gouging, ripping. "Some people call me Badger." He's prospected in the Solomon Islands, Vanuatu, Indonesia, Panama and the Kalgoorlie area in Australia. In most of these places, he says, "a rock is a rock; copper, lead, zinc alteration minerals are universal."

Most of Bell's time, though, has been in the Coast Mountains. "Because ore bodies are small you have to get your head into it; you can't go too fast. If you took a sample of everything that looked favourable in the Coast Range—fuck; your pack would be a thousand pounds." He drank from his coffee cup like a beer drinker swilling a pint, and muttered to himself: "All this lead around. Does it always have to be galena?"

He came to prospecting via an interest in the outdoors. He was raised in Jasper, where his father was a minister at the St. George Anglican Church. His mother was a nurse with a passion for nature. Bell got into hiking, then climbing. With friends he scaled Pyramid Mountain, Old Man Mountain, the North Boundary Trail. Then he began solo jaunts: the South Boundary Trail, two weeks with his thoughts and his fishing pole bungeed to his pack. In the mid-1970s and out of school, he arranged to take over a trapline at an uncle's ranch in Bridge River, BC. He rebuilt the cabins, honed his bush skills. Then, as word spread of his competence, he took on work as a range rider. He looked after cattle from nine ranches, packed salt, kept the herd moving to new ground, watched for predators. There were six hundred cattle in all—Hereford-Shorthorn, Red Angus, Black Angus. In the winter he trapped marten and mink. One time he came off the trapline and heard an unfamiliar sound. It was a human voice. He hadn't heard a voice for four months.

"Why did I do it?" he asked himself. "I thought: 'I'm doin' the bushman thing.' I'm really proud that I could do that." He forked more egg into his mouth. And added, "Maybe I was just running away, too."

A combination of remoteness and cheap land lured Bell to the Kispiox Valley, where he bought one hundred acres. The property was fifteen miles up the valley, beyond power and much of a road for that matter. He was wintering in his cabin when a former Geological Survey of Canada mapper named

Tom Richards, who had moved to the valley in 1977, began to hold classes. Richards' idea, according to Bell, was to form "a syndicate of the bush-wise." Richards would teach them basic prospecting skills and they'd fan across the region. Thus was born STROB: Swami Tommy's Revolving Ore Bodies—a hallucinatory name in some way connected to the substance-use habits of some members. Membership included Pat Suratt, Ray Cournoyer, Paul Huel, Bruce Holden. They held classes in people's unfinished houses, sitting around homemade tables illuminated by a mechanic's trouble light knotted from the rafters. Or they used the venerable Kispiox Community Hall, teeth chattering until the heater kicked in. Richards would arrive with his buckets of rocks: sphalerite, galena, feldspar, barite, quartz and dozens more. They'd take home fifteen to twenty rocks to study during the week, then be quizzed on them the next time they met. "Tom was very patient: We'd say, 'What's this again?' He'd tell the whole story."

With a growing reputation as a first-rate prospector, Bell worked for ten to twelve years in the Coast Range for Ralph Shearing, then for Henry Awmack and Dave Caulfield at Equity Engineering. His first independent job was for Reg Davis on Johnny Mountain where the Snip Mine was later developed. He worked in the summer, sunk some of his savings into a sling of plywood for his Kispiox cabin, and then jetted south to fish steelhead in the rivers of Patagonia. When he started, the rate for a prospector was $150 per day. Now it is $250 to $300. A bigger change is in attitude. "The big companies don't believe in prospectors anymore. We're just dumb dirt baggers as far as they are concerned. They can't believe it when you come in with a juicy rock. The geologist is incredulous."

He told me about a particularly satisfying find he made on

the Fawn property in the Nechako Plateau that, despite having been trenched and examined several times, had befuddled geologists. Soil geochemical surveys had suggested an ore body on the property, yet trenching and testing had always failed to identify it. Digging around in the roots of an upturned tree, Bell found barite, a mineral often associated with gold. The previous company's geologists had walked right by it. The barite focused Bell's efforts and he found a showing nearby. The geologists got reinvolved and found that the ore body did not cross the trenches as they had previously supposed, but actually paralleled it.

Before Bell and Diane got up from the table I asked Bell what he'd do if he were given one hundred thousand dollars for exploration. He said he'd go back to the place in the Whitesail Range where he and Tom Richards worked in the first two years. They were looking for a source of placer gold. "I've still got my diaries," Bell said. "I didn't know shit from honey."

After just three days I hadn't slipped into the easy come-what-may lope that, I think, makes for the best notes. I was rammy. I was irritated at Tom Bell and Diane for not taking me with them for the day, for not driving me over to another prospector's house and saying, "Here, tell this guy your best story." I wasn't relaxed. What I needed was mainline information.

And that is exactly what I found in the records of the Gold Commissioner's office in Smithers. The office is located in one of those incredibly useful small-town government services centres where, in one stop, you can get a driver's licence and a permit to dump industrial waste, and report your neighbour for cruelty to a horse.

Don McMillan, the Gold Commissioner, was short and

sturdy with the pallid skin of a man who has toiled for decades under fluorescent lights. His right hand seemed in a permanent clench the size of a Tim Hortons paper coffee cup. As Gold Commissioner it was McMillan's job to issue Free Miner certificates. The task, he said, is not onerous now, but there have been times when things got complicated. Some years ago the citizens of the nearby settlement of Round Lake came to him in a tizzy after they had woken one morning to see trees in their front yard sporting axe blazes, the thin string of hip chains webbing across their property and mineral claim posts driven through the heart of their azalea gardens. After some investigation, it turned out that overzealous prospectors from Saskatchewan had staked the area. The confrontations, McMillan explained, were caused by the free miners' secrecy and the property owners, who "see 'an open-pit mine in my backyard' and panic."

McMillan is guardian of a repository of mining records dating from the late 1800s. They are kept in a small room off the main office. The walls are lined with big books. Heavy as a watermelon and bound in scrofulous red leather, each volume included the details of mineral exploration for the area for more than a century. I grabbed a random volume, flipped it open. It smelled of rock dust. Omineca Mining District, fall, 1904. The writing was copperplate. Harry Howsen staked a claim on Howsen Creek on September 3. He called the claim Evening. I flipped through the pages, records of trails cut, trenching. Claim names with promise: King, Rain Bow, Hunter, Waresco, Ethel, Cracker Jack, Lucky Jim, Silver Tip, Reliance, Perhaps, Enterprise. The entry for the Prince of Copper claim is typical: "trials cut open. Open cut 15 feet long, 10 feet wide, 12 feet in height." Then a heartbreaking note: "Solid rock." The journals are a notation of hard work in depth and width. Heather Bell,

Copper King, Lukens, Virginia, Copper Queen, Kyle, Tenderfoot #1.

Bonanza has always been a popular name for claims. In one of the heavier books in the Gold Commissioner's office, it appears as a notation for the year 1906, when it was first staked by the enigmatic James Cronin who, I was later to learn, is perhaps the greatest figure in northern BC mineral exploration.

By the time James Cronin arrived in the Bulkley Valley in 1906, he had a reputation as a one-man force of economic invigoration. Born in Ireland, he had apprenticed in Nevada silver mines, learning how to prospect and relate mineralization with rock types. A strong, energetic man, he could timber a mine, sink a shaft, draft and engineer. What separated Cronin from many others, however, was his ability to wick the value of an ore body from scanty evidence. Just by clutching a rock he could accurately estimate its assay, and after several hours in front of a rock face he could pronounce it worthy or not of mining.

In the 1890s he stopped in at the St. Eugene Mission in the Kootenays and while talking with a Father Coccola he noticed some odd-looking rocks strewn about. The Father explained that they had been brought in by an Indian named Peter. Cronin was taken to the showing and staked the area. Developed from that simple observation, the St. Eugene mine would become one of the biggest lead and silver mines of its era in BC. Nor was Cronin finished in the Kootenays. Nearby he noted the possibilities in the working of two abandoned mines; one, the War Eagle, went on to produce millions of dollars of ore over eighteen years.

Next, Cronin turned his attention to the largely unexplored northwest corner of the province. Travelling by foot, horse and canoe, he surveyed the area around Hazelton and settled his considerable attention on a small property forty-five kilometres southeast of Smithers in the Babine Mountains. Called the Babine–Bonanza claims, it had been located by two prospectors the year before. For the next fifteen years, the Babine–Bonanza became the focus of Cronin's efforts.

From the very start Cronin's instincts told him this was a property worth mining, yet the obstacles to it ever being a mine were great. The property was walled behind towering mountains and moated by an infinity of turbulent and as yet unnamed streams. The only access to the site was via the Moricetown trail, a tortuous route of switchbacks and gullies plagued in summer by swarms of flies. When the railway created Smithers in 1913, the Driftwood Trail was built, which cut an impressive thirty kilometres from the trip. At one point the trail rises 500 metres in 1,200 metres.

It was lugging supplies into his claim that Cronin conceived of one of the stranger methods of transporting ore: by Zeppelin. By 1917, with the war in Europe reaching a zenith, demand for minerals was high. Cronin, who had heard that German Zeppelins had been captured, wrote to then BC premier Sir Richard McBride, saying, "Transportation of ore was the only thing these machines were good for." Unfortunately, McBride's response is lost to history. Yet despite the obstacles Cronin was so convinced of the potential of the property that he willingly spent his fortune developing it. He later wrote, "The opening of this mine has proved to be the most tedious undertaking of my life, and having staked everything on the finding of a paying mine, the undertaking is also the most serious. I feel quite determined to stay with it until the full value of the property is fairly determined." And again, four years later: "To win in mining, the game must be played to a finish and this we are now trying to do."

Finally, on a fall day in 1923, with the scarlet in the birches and the Vs of Canada geese winging south overhead signalling the end of another season, Cronin, now an old man, bent and largely beat, decided to abandon his claims. He felt he had proved the property capable of supporting a hundred-ton-per-day mill yet he still had no takers. He and several employees had been trenching across a zone, hoping to provide further evidence of mineralization. Just as dusk settled on the camp they let off a final blast, loosening rocks for the next day's excavation.

The next morning Cronin woke uncharacteristically forlorn. While the others slept he walked into the workings and kicked at the fragmented rock. It shone. Incredulous, he overturned chunk after chunk, each time met with glistening ore. Had he discovered a major ore body? He shouted to the others and they ran with shovels and picks at the ready. Alas, their jubilation was short-lived. Behind the sliver of mineralized rock was a face of gangue—worthless rock.

The find didn't change Cronin's mind about continuing—he was set on leaving—but it did make real the possibility that he could off-load the property for a profit. Carefully, and like a stonemason, he reassembled the chunks into the hanging wall, creating the appearance of a major ore body in the face. Any greenhorn speculator, of which there were many in the 1920s, would snap it up for a tidy price. But Cronin's bad luck hadn't changed. It was just on hold. It happened that a government district engineer was in the area and came to the property for an inspection. He took one look at Cronin's reassembled trench and leapt in, pick in hand. A few hours of picking and the entire façade was rubble at their feet.

As close to crushed as he would ever admit, Cronin expressed his feeling in a letter a few days later. "I could see $50,000 disappear with his pick work and I shall always remember that it was the greatest effort of my life to hold myself [from] using the pick on the man's head. In a few minutes all my prospects were spoiled . . ."

To get to Pat Suratt's home in the Kispiox Valley, I turned right off the highway in Hazelton, passed over the Bulkley River on an insane suspension bridge, then sped down a winding paved road that turned into ball-bearing gravel. Clutches of kids and dogs stood at the roadside. If the dogs weren't white they were black. The trees were powdered in road dust. Several kilometres past a knot of plain houses I turned right onto a tree-lined lane and followed that to the log home by the river. That was not Pat Suratt's place. But from the terse manner in which the residents greeted me, I understood that Suratt's visitors often mistakenly came that way. They were his unofficial doormen.

Suratt's was the next driveway up, which was blocked by an all-wood swing gate. It was clasped with a keyless lock that perplexed me for fifteen minutes. Used to getting my own way, fast, I returned to my truck and was going to drive to the neighbours' to plead for help when I recalled seeing a similar lock in a book on gates. There was a trick to it. I fiddled and eventually triumphed. Humbled, I drove in, descending a road to perhaps the most stunning rural property I've ever seen. The driveway snaked past a scrub of forest on the right; a pasture gave way to a fringe of trees through which the Kispiox River glistened silver. The farm itself was on a bench, a rich flat of pasture in the middle of which stood an ancient horse. The woodshed was full and it was nicer than the house, though that is not such an awful thing as the woodshed was very nice. The property was low, and all the buildings had the welcome languorous feeling of a slept-in bed.

Pat Suratt opened the door with a towel clasped in his hand. The smell of steamed salmon rolled around him. He was canning sockeye. They were stacked in the tub on his sink like firewood, enormous orange slabs. The kitchen was rank with brine, the smell of hot Mason jars. He said he could talk but he had to keep on canning. He'd be at it until midnight.

Suratt was of that subset of sinewy male that only the Americans seem to be able to produce: powerful forearms, tall, cordy; appearing genetically predisposed to swing a pick or heft a mattock. He spoke with a slight drawl.

The kitchen was lined with wood planks, and on the windowsills were rock samples: bull quartz, pyrite, and a fossil of what Suratt described as "a squid-like character" from a dig near Telkwa. I sat at a spruce table that he'd made soon after he bought the property in 1975. He'd made the table from parts of a cabin that had been there since the 1930s, when a Panamanian sailor had purchased the property via an ad in an American newspaper. Apparently the ad boasted of riverside property in BC. He paid three hundred dollars total. He put fifty dollars down and paid it off at five dollars per month. He built the cabin and left. After Suratt bought it he added onto the place, fenced it, created a paddock and a huge garden with a wooden fence like a hem. The vegetable garden used to be on a bench below the house and next to the river. But one year the river spilled its banks and the garden washed away.

The table was good and solid, and I laid my notepad out. There were two books on it: *The Joy of Mathematics* and a novel by André Maurois. I asked Suratt about Tom Richards.

It was about 1978, he recalled, and Tom got a night course going. Pat's voice was bourbon and woodsmoke. "There were fifteen students. There was no charge for the course. We'd meet at someone's house, drink beer, smoke dope, look at rocks." He said the group had a reputation as crazy hippies. Many wore ponytails. There were hangers-on, and hangers-on of the hangers-on. When they knew chert from granite, Richards took them on an early field trip out of Cedarville, where they camped in a spectacular daisy meadow. They had trooped in wearing ill-fitting five-and-dime-store Czechoslovakian-made packs. The liquor came out, then a shotgun. Suratt said to the group: "You are making me nervous."

The next day they prospected. Fanning out, they worked their way up a draw, and then paused by a waterfall to have a joint. Below them the landscape formed a bowl, as smooth as a blanket on a sleeping body. They noticed one of the hangers-on, a heavy-set former Ontario taxi driver named Mike Clarke struggling through the bush far below them, waving. They waved back. What they didn't know was that in Clarke's pack was eulachon, nor that as he walked down the berry-laden path that he'd felt a presence behind him, and turned only to come face to face with a grizzly. Clarke later said that he thought it remarkable how brave they were.

On the bank of the creek, the group was calm until the grizzly turned its attention to them. It was maybe three hundred pounds. Cornered in the narrow creek, with the waterfall on one side, the group's only option was to scramble one at a time, up the bank. The bear tried to exit the creek, but it could not get up the slippery clay bank either. Each time it failed it got more frustrated. Angered, it swatted rocks. It hit a boulder and sent it flying.

In their panic to get out the men got careless. Tom Richards dislodged a boulder and it tumbled down, narrowly missing Suratt. Suratt was the last one up. By this time the bear was in the creek and very close. Circling back to check on Clarke, they found him in a tree. He had peed himself, and was convinced the future lay in cabbing, not prospecting. It took some effort, Suratt said, to talk him out of the tree.

Eventually, and under Tom Richards' tutelage, Suratt became a first-rate prospector. During the 1980s and 1990s his income for working from May to October was as much as thirty thousand dollars per year. He was one of the elite: a prospector who could earn a living. He worked in Gold Bridge, Nechako, Chilcotin. On Blunt Mountain in the mid-1980s, Suratt and Richards were working on a joint venture project with Noranda. A previous crew had studied the area without coming up with an explanation for its geochemical anomalies. They were called in to solve the riddle. Exploring the property, Suratt found an area where the camp had been set up. With his nose to the ground like a hound, he followed the path the crew had taken to fetch water. He found float—rock fragments—of almost pure galena. They had walked past it while doing their daily chores. Then, further on, he found a foot-wide vein of galena. The crew had stepped on it on the way to fetch drinking water.

Suratt said he had a system for prospecting. Before working an area each day, he'd have the chopper pilot fly over it. He said that one time, cloud prevented the helicopter pilot from doing this. So the chopper set him down in a creek. His plan was to walk the creek out. But partway through the day Suratt found that the creek, as he put it, "steeped out." Faced with a choice of walking back the way he came and proceeding, perhaps facing a night in the cold, he opted for the latter. He inched his way up a sedimentary ridge. The footholds were shitting out. For a while he thought of waiting for the helicopter. That's the ultimate: to be plucked from a ridge. But he calmed down and saw that if he kept his head about him, there might be a way out. He inched ahead. It took him two hours before he was off. He had gone two hundred metres.

Suratt has no time for promoters and for that part of the exploration industry prone to exaggeration. He says that the way the industry used the loss of the mineral property at Windy Craggy, on the Tatshenshini River, wasn't straight up. "The way the industry portrayed it was a matter of fact that it would have become a mine. But copper was at sixty-seven cents." At that rate, he said, it wouldn't have been economical to mine. He calls these sorts of mining stories "folklore." He cited another one: When Tweedsmuir Park was created, Deerholme mine was forced to close. But the talk of its riches not only persisted but also grew. There was talk of a ten- to twelve-foot-wide vein. It had potential for four hundred thousand tons of gold-bearing ore. There was a fancy geological theory that the ore body in the mine was somehow the focus of an epithermal system that ran from the Whitesail Range clear to the coast. Then the government rejigged the rules so the mine could be reopened. Suddenly, said Suratt, the exploration companies weren't interested. "It wasn't even worth bidding on."

I left Suratt and his hot, briny jars and drove to the Petro-Can at Kitwanga. The parking lot was the size of a Vancouver Island paddock. It was so big, in fact, that I puzzled where to park. The attached restaurant was full of truckers and Indians and seemed to have every possible type of bad food, as well as freshwater fishing gear. I bought a Coke and salt and vinegar potato chips.

Something happens to distance after Kitwanga—a lengthening similar to what happened to terrains represented on Mercator maps. It is a feeling like being on the water on a slow boat. You look at the land and wait for it to change but it never does. The mountains huddled on both sides of the road as if cold. The sky was the colour of an aluminum pot.

Stewart, once known for its all-night dances and marvellous fist fights, is now dead. I checked into the King Edward Hotel, a blocky chunk of mediocrity that, in most towns, would have burned under suspicious circumstances years ago. The woman behind the counter was as friendly as a cardboard cutout. But it was late at night and she had me. There was no other place to stay.

But the town had her too. A skinny, hollow-eyed woman, she had the post-glamour looks that reminded me of X.J. Kennedy's line about the former beauty who has "two toadstools for tits and a face full of weeds." As she fiddled with the bill, I played my favourite hard-luck game: What Went Wrong. The suburban girl with the inflated sense of beauty dumped a course in medical stenography for the transient delights of a curly-haired assistant manager in a Trans Am. Then: male pattern baldness, a Bino's on the corner opposite, and before long they were stumbling around north BC with enough money from her mother to buy into a seasonal hotel. She's too disillusioned to even try and get a lift out of men.

I unpacked and set out to find dinner. In the trendy Bitter Creek Cafe the waitress greeted me like an old friend, then forgot to serve me. The restaurant was almost empty. Beside me four Americans from the neighbouring town of Hyder, Alaska, sat at a table heaped with food. The old man, bedecked in a ridiculous blue plaid shirt, US stars-and-stripes hat and stars-

and-stripes suspenders, mumbled: "Where are we?" The son-in-law, sporting a revolutionary holiday five o'clock shadow, said, "British Columbia." There was a pain on the old man's face. I bet he hasn't had a good time since the Korean War. His lips trembled. He said, "British?"

When I left the King Edward at 4 a.m., the night clerk, a pasty fellow in his early thirties, looked up from his book and grunted so-long. The book was *Descent Into Madness*.

At 5 a.m. on that cold Wednesday morning I really, really wished the defroster in my truck worked. For three hours I drove, huddled over the steering wheel like a crazy man. If I had the window up, the window fogged. If I put the window down, my ear and neck froze. When I finally did stop, at a swish and totally remote resort named Bell II, I was so stiff and bandy-legged that I had to walk around the parking lot to work out the knots. Bell II was a strange place, built on what looked to be a clear-cut, with big fuel tanks but otherwise done up to please a wealthy Bavarian BMW executive come for heli-skiing. There were rock walls around the gas pumps. Very tony. I had a big breakfast, filled the truck's nearly-empty tank and then . . . realized there was no bank machine. It was my first sense of being a coddled semi-urbanite. Caught without cash. With the twenty-dollar bill in my wallet and handfuls of change retrieved from the ashtray of my truck, I just managed to pay.

The entrance to Eskay Creek Mine, one of the richest mines in Canada, was up the road a few miles. It was barred with a hefty yellow gate. I fumbled it open with a key for which I had more or less signed my life away several days before in Smithers. I was headed up this company road to kilometre 45.5. To make sure that I didn't head-on with a loaded ore truck, I had outfitted my truck with a portable two-way radio rented to me by an outfit in Smithers. The instructions given to me with the gate key were to radio my position every two kilometres. When an oncoming truck reported a distance close ahead I was to pull into a wide spot and wait for it to go by.

With every kilometre the forest and surroundings looked more coastal. The pines gave way to some hemlock and stringers of lichen trailed from the branches. The underbrush was rank and, it appeared, terminally wet. The road, built to access the mine, crossed the rain-swollen rivers and wound along mountainsides. The kilometre markers were tall so they could still be visible when thirty-five feet of snow had fallen. The radio cackled. A truck was several kilometres ahead. I quickly pulled over. Minutes later a fuel tanker rumbled by. The driver, who was eating lunch, waved a sandwich in greeting.

I found what I was looking for on the left side of the road, just past a sturdy wooden bridge. A settlement in miniature, it centred around four trailers set side by side, several large camp tents, and a garbage incinerator from which a nasty-smelling black smudge emerged.

There was a clatter and a curse from one of the trailers and a small, sinewy man clad in an apron that reached from shoulders to knees leaned from the door. He spotted me, curled his index finger in a bossy way, and said, in a manner completely at odds with the industrial clutter, "Would you be so kind as to give me a hand?"

I stomped across the wooden rounds that had been set in the mud slurry that surrounded the buildings, up the stairs and into the trailer. It was as big as one of the medium-sized motorhomes you see on the road, and divided in two: commercial kitchen in one half, seating and lunch counter in the other half. It was a kind of miniature restaurant, something like the Smithsonian might display as an example of twentieth-century resource exploration culture. Three white tables, clutch of condiments at one end, plastic chairs. Everything was designed for wipe-down, put-away ease.

The cook, Will Elliot, handed me a big flipper. He was roasting chicken legs and thighs, and wanted to drain the fat into an enormous caldron of soup. He had tried by himself and the legs slid out of the pan. I placed the flipper at one end and he poured. "Beautiful, beautiful," he said. It was the first truly useful thing I had done in days. He rewarded me with an offer of soup and coffee.

Elliot was slight with a deeply creased face under a fop of black hair. Dark lines around his eyes made it look as if he'd been wearing a scuba mask. He looked like an old young guy. He was in his mid-fifties but when I asked about his past he dated himself to a time in the 1970s when he got interested in cooking. Apparently, he had a set of personal circumstances that precluded discussion of his earlier life.

Elliot told me he was on the last camp for the summer. He had started in the spring at a diamond camp in Aylmer Lake, then moved to a camp run by an outfit he called the "two Daves" somewhere in the north and finally to Eskay Creek. He said that the crew at Eskay was one of the best he'd worked with, and included such characters as Brent the Brain, Sean the Destroyer, Nervous Dave, Kiwi Scott, Mr. Five Reasons, Mr. Hélicopitère and, somewhat ominously, Commander Woo. He called this crew the Heroes of Labour. I spooned up his soup—a remarkable mix of sausage and vegetables—and asked how he got into cooking.

He said he started cooking in the West Kootenays, in an era that he describes as "a little before Elvis Presley died." He was working as a chainsaw carpenter, building retaining walls on logging-road bridges "so the trucks wouldn't fall off." He saw a posting in the cookhouse. He had never cooked before. "I said, 'Give me a shot. Maybe I'll cut the mustard. Or I won't.'" The cook was crabby; he had fired many helpers. But Will was optimistic. He lasted three weeks.

His cooking career might have been over but for the fact that he had a first aid ticket. Camps of a certain size must have someone with qualified first aid. Elliot said the camp cook must be adept at four things: cooking, bottle-washing, baking and first aid. Most injuries are "vegetative"—maybe a cut finger. Now he picks and chooses jobs; in the spring of 2002 he received five offers in one day.

He's worked in logging camps and oil exploration camps. Mineral exploration camps are more democratic, he says, than logging or oil exploration. The distinctions between the top geologist and the lowest dirt bagger are less obvious than they are between a faller and a chokerman.

Elliot cooks out of recipes in his head. Early in his career he decided, he said, "to strengthen my repertoire with lasagna." The recipe called for spinach, tomatoes, three cheeses and wine. "It was fucking deadly," he said. He laid out the dish on the serving counter. A driller led the dinner charge that night. He had a plate in one hand, a fork in the other. He paused at Will's great lasagna creation and snarled, "What the fuck is this ragtop shit?"

Elliot's rules for cooks are to not run out of (in order): coffee, sugar, flour and rice. His ambition is to make the cookhouse free of strife. He gets up every morning at 4 a.m. and paws through the seventy-five to eighty CDs he brings into camp. If the morning is lousy, as it was on this morning, he'll put on k.d. lang's *Shadowland*. "She sings as if she's down to her last cigarette," he says. But he bakes to opera. "I don't understand this opera. I don't get the libretto. But the emotion is there."

A big redheaded man stomped into the trailer. He wheezed, breathing through an open mouth, glared at me and said, "Who are you?" He had the broad span of a man who has leaned over map tables and studied rock samples. His glasses were thick and, when he took them off, revealed heavy creases around his eyes. Here, I thought, was the head geologist. And I was sitting in his cookhouse. I didn't want to go back out in the rain. So I mumbled something about waiting for a chopper. He stared at me long and hard as if working through the costs of me slurping his coffee, then walked into the kitchen, lashed an apron around his girth, and squirted a great gob of dishwashing soap into the sink. Pete was the dishwasher.

Elliot and I talked for another half-hour until we heard the distant pounding. The cook set down his ladle. "That's your ride," he said.

I stepped outside in time to see a sleek little bubble of a helicopter roaring over the clearing, drift sideways—like a car in four-wheel drift—then set down with the finesse of a hummingbird alighting on a thorn bush. I was bound for the Adam property near Eskay Creek, where I would help Awmack and his group move camp and then spend the next four or five days accompanying them on exploration of an intriguing area rich in possibilities for copper and gold.

The pilot's name was Lenarduzi. He wore dark blue pilot coveralls. The helicopter was a new Hughes 500, often described as the sports car of helicopters. Lenarduzi went over some safety issues. Tail rotor, safety beacon, how to get in and out of the aircraft. "Best not to get your head over the bubble," he said. Apparently the idea was to duck on the way in. In Hazelton a pilot was cleaning the bubble when he absent-mindedly put his head up. Even though the blades were rotating to a stop, the impact killed him. My attention was cut short by the engineer, who was loading a cargo that included boxes of fresh lettuce and fruit, as well as jerry cans of diesel and gas. These were heaped as an ill-fitting jigsaw on the seat from floor to roof. In a flash of daydream, I smelled half-burnt fuel, the stench of burned skin and a top note of scorched chard. A small cell-phone-like package of bear spray was strapped to the front strut. If a canister goes off in the bubble chances are very high that the aircraft will go down in a heap.

As well as the generic hazards of helicopters, each type of craft had its own dangers. The sturdy JetRangers, for example, have long sagging blades that required a person entering or departing to take extra precaution to duck. With the Hughes 500, the risk was from the superheated turbines. Apparently the novice passenger steps behind the craft, gets a blast of superheated exhaust and stumbles into the tail rotor. As one geologist cautioned me, "It's a stir-fry kind of thing."

We rose in a surge of power. The sun was hidden behind a turf of off-white cloud. The land in the Iskut Valley is rough in a tattery sense, not in the toothy glamour of the Rockies. The valley bottoms are heavily treed, flat with a river of brown water and white froth coursing through. From the valley bottoms the sides rise into hills and the hills into mountains. I was immediately glad that I was over and not in that country. We flew all the while west toward the US border, climbing as we went.

Lenarduzi said that we would try to get into the Adam property, but the clouds were thick and he couldn't be sure we'd be able to set down. I felt like we were flying under an upside-down bowl of mashed potatoes. We circled around a great white duvet of cloud, rose and swooped, and emerged over a pocket valley between two red-stained peaks. To my right was the larger peak, with an almost gaudy-coloured gossan canyon rising to a small bench land, and, fronting it on 180 degrees like an amphitheatre, a slope of talus—broken rock fallen from the cliff above—and outcrop. A fringe of forest below the canyon began and swept down to a boulder-strewn creek. I glanced at my map. King Creek. King Creek did not meander, to use the geologist's term, but instead rushed headlong down the valley.

After fifteen minutes Lenarduzi found an opening and we swept across the valley and briefly circled a little camp. Two tents were set against a slope. Around them was scattered a litter of generators, packs and jerry cans. It looked neat and cozy, as if you could pick up the tent with your fingers and move it somewhere else. We circled again. Then Lenarduzi set us on the rocks and since he was staying, shut the machine down.

There were three people in the camp: Henry Awmack, the geologist who arranged the trip; Tim Sullivan, a prospector; and a student who, because of his unhappy circumstances, I'll call Avie. There was a hushed, awkward feeling about the camp which I couldn't identify. The pilot and I entered the tent and were immediately confronted by the smell of unburnt fuel, unwashed male bodies and fried food. The tent was cold because they had disconnected the stove in preparation for moving. The three of them had a slightly shell-shocked look, in part, we discovered, because they had weathered a week of awful weather: snow, rain, constant winds. In some twenty years of exploration geology, Awmack said, he had rarely been unable to work. Yet one day it blew so hard that they didn't dare venture from the tent. Sullivan, short, sturdy, in his early thirties, pointed to the tent's metal supports. As thick as my forearm, they were bent with the force of the gale.

The idea was that I would help them move, and then spend the next four to five days accompanying them on basic mineral exploration. First Lenarduzi would fly Awmack to the new site with the load of supplies, where he would pick out a campsite. Then the chopper would sling loads of camp gear that Tim and I had stowed. Since the day was half-over there was no time for chatting.

There was a roar as the helicopter left, and Sullivan went to dismantle the generator. Avie and I began to pack supplies. He was in his early twenties with curly hair and a clear smooth complexion. Raised in Grand Forks, he had gone to the University of British Columbia to study geology. He told me that the idea of working outside in the mountains had attracted him, so he had pursued economic geology. This was his first field trip and, he said, maybe his last. Many things had gone wrong. The elements had conspired against him. Howling winds had kept him awake and he was overtired. Wind and snow had frozen his fingers. At one point, working alone, he had encountered a bear. He tried to radio the others but discovered that his antenna had vanished. At one point he'd actually wept. "I don't know how these guys do it," he told me. "But I can't handle it." He was flying out with Lenarduzi. Whether he was done with mineral exploration or not he didn't know.

I wanted to give Avie some game advice of the buck-up-it-will-get-better variety but was hardly in a position to do so since I was feeling slightly overwhelmed myself. In the half-hour since I'd left the comforts of Elliot's kitchen, I had a) arrived at a windswept, desolate place with no reference points; b) been accosted by bad smells; and c) was getting my feet wet. In the days to come, when I fetched over a ledge on my own or clung from fingertips on a rock bluff, I had similar feelings. It was a kind of existential "What the hell am I doing?" As I discovered, the hardships are not really so bad, but what was difficult was the transition from one situation—comfortable, familiar—to the other—harsh, cold, foreign. Avie, who very much looked as though he'd been raised in a warm, loving home, had simply been overwhelmed by the difference.

The site Awmack picked for the next camp was a small rise below a talus slope. He marked the camp, as he always does, with a Canadian flag. Two tarns were nearby, a small one which Sullivan decreed would be for drinking, and the larger for washing. The gear was still in the nets. It had started to rain, so I donned rain gear. Sullivan watched me, then said, "Well, you're here to learn. You might as well do something." There was something of the tough-guy in him, garnished with Commercial Drive coffee shop modernity. Setting up a camp is a matter of a thousand little decisions—from views to sewage control. A tent set upon a knoll is likely to be free of mosquitoes and bugs, yet vulnerable to storm. There needs to be water nearby but the tent can't be on a flat that might flood. The tents are self-supporting, but the stovepipes require wooden poles. If the camp is above the tree line someone has to take an axe and fetch poles. All the time we worked, there were these and many more decisions to make. Which way to have the tent flap open? How to configure the inside of the tent? Where to dig the crapper?

Geologists are like those famous Dutch farmwives who, by the look of the farmhouse floor, assemble a whole belief system around the way someone lives. Sullivan decided—sensibly, I thought—to put the stove in the kitchen tent. That meant the sleeping tent was unheated. A warm kitchen seemed a good thing. Yet, when I later mentioned this configuration to another geologist, he said, "That's crazy. I don't know why he does that."

After days of driving and note-taking, it was good to work with my hands. We assembled the skeleton of the tent—again, in a deliberate way; otherwise there was grief later in the process—then pulled the skin of the tent over it. It was like skinning an animal in the opposite direction. Then Sullivan, weary of having to tutor another greenhorn—this one with a notebook—set me to loading the lip of the tents with boulders.

It took two hours start to finish to complete the camp. What was erected was fully capable of sheltering the three of us. Two tents, one with a ticking hot oil stove and kitchen table, the other with fold-out cots set up and our sleeping bags laid out. We could have lived on the mountaintop for years. Awmack made spaghetti and a large salad. As a final duty, I was to take a large red and white cooler filled with meat and bury it in a fringe of snow that remained by the larger tarn. The fringe was hard, and I stabbed at it as I would at clay. Eventually I broke through. I set the cooler in it, covered it with snow. I paused and looked at the camp across the little lake. The water was completely rippleless and reflected the tents so there were four. It was too dark to see the mountain, but I could feel it above and around me. I munched on a snowball. It had been a long day.

To be woken by an alarm clock while sleeping in a tent at the top of a mountain is a surreal experience, but something had to get us out of our slumbers, didn't it? We—geologist Henry Awmack and prospector Tim Sullivan—were here to work. They stirred in their big sleeping bags, rustling and making those waking sounds that people do. Then they dressed and started their day.

First, Sullivan walked twenty paces over the hillock and pulled on the starter cord of the little Honda generator. It was as quiet as a refrigerator and set the lights blazing. Then he unzipped the fly of the kitchen tent and rustled in a plastic tote of pots until he assembled the parts of a percolator coffee maker. This seemed hard for him; Sullivan has a hard time with mornings. He blinked at the heathered floor, as if trying out a new pair of eyes. Only when the tent was redolent with the smell of coffee and he had poured himself a cup did he say anything. And that was, "Yeah."

Awmack, meanwhile, was into maps. He didn't look awake either, but he was keen and he wanted to have the maps in front of him when he came to. Sitting there, map held limply in front of him, he looked like a man in a boat waiting for a breeze to fill the sails. The map was of a series and imposed on the contours some geochemistry. A coffee and a refill and he was thinking aloud and addressing no one. "There is some disagreement on the origin of the rocks. Most previous studies say they are monzonitic, but one group says they are extrusive volcanics. The truth is important to the . . . it makes a difference on whether the Eskay story were possible." If it is possible, then Awmack's clients, who have the claims, have an incredible story to tell investors, and very good reason to carry on with exploration. A world-class gold mine, Eskay Creek produces millions of dollars a year in profits. When Eskay's miners are working in particularly high-grade sections of the mine, a single truckload of concentrate contains as much as a million dollars' worth of gold.

While Awmack was doing this, Sullivan made breakfast and his lunch, and then while Awmack made his lunch and did the dishes, Sullivan assembled his gear. They had obviously worked this way for a long time, one doing a sort of household/domestic thing while the other did a work thing. They slipped effortlessly back and forth; by eight o'clock, the camp was tidy and clean and they were suited and staring at the mountain above them.

Standing there looking at the mountain, Awmack attempted to separate what he had read about the property from what he saw and, truth be told, what he wanted to see. All three—report, rock and hope—were languages of sorts, and equally capable of deception. His challenge for the day, as it is for all exploration geologists, was to read the rocks in their own words.

Ahead of us was a continuum, a rising staircase of outcrops veiled in talus, of rock from basalt to rhyolite—a rock suggestive of deep ocean ducts and naturally condensed ores. We started thirty paces from the tent. The flag waved southeast. Awmack banged an outcrop, glassed it with a small magnifying glass called a loupe that he wore on a string around his neck, and said he thought he'd found dacite. This was good. Dacite indicates a shallower area than monzonite. The kind of deposit he was looking for would have been vented from a crack in the earth's crust in less than a thousand metres of water. To find dacite here meant we were in shallow water.

Mineral exploration is about patterns and analogies. No two mineral deposits are exactly alike but there are enough commonalities among deposits to recognize families: epithermal, skarn, porphyry. A geologist will look at a characteristic and say to himself: this suggests a certain kind of mineralization system might be present. Rocks and rock formations come in so many variations that geologists have developed models out of necessity. They walk about in the field with a model in mind, testing a hypothesis. Does what they see support the model? If not, then what does it support?

The danger in this kind of thinking is that a geologist may only see what he wants to see, and in doing so risks ignoring indications of another model. The most famous example of this is the Canada Tungsten Mine on the border between the Yukon and the Northwest Territories. The area was first explored by Kennecott, which was looking for copper. When they didn't find enough copper they left. Another company came along and, in analyzing Kennecott's samples, put a black light on the rocks. The light revealed tungsten. "You won't find what you aren't looking for," is a common motto in exploration geology.

For this reason, an area can never be said to have been totally explored. Another geologist looks at a property from a different angle and sees different possibilities. Or a new deposit is discovered elsewhere with some interesting features in common.

Just a few miles from us, the Eskay Creek mine was a reminder of the risks of sticking too rigorously to one theory. The area had been thought to have great potential since the 1930s yet no one was able to unlock its riches until the 1990s. Why? Because the ore body at Eskay Creek does not outcrop. Only veins appear at the surface. Geologists understandably sought to understand the deposit in terms of veins. Yet it was only when another model was applied—of an ore body lying in the mud on the flanks of an underwater volcano—that the nature of the deposit became clear.

A few paces later, we were in monzonite. The explanation? Maybe a dyke. A dyke is the filling in a crack. It can carry another rock—like a subway line—far into the host. Monzonite is not good. Its large crystals suggest underground cooling—and support for the low-grade porphyry model, which Awmack did not want to find. We wanted fine-grained rock that was disgorged into sea water and cooled quickly. We passed grey, foreign-looking float and Awmack hardly gave it a glance. Andesite, he said. He looked up and a tower of grey rock loomed overhead. It probably came from up there. Tracking float was Sullivan's job. Awmack wanted to lay out the geology, the architecture of the mountain.

I banged a rock too. I was outfitted with an Estwing rock hammer, a loupe, plus an assortment of compasses, notebooks and an emergency kit, all tucked into a very handy multipocket vest. At one time the vest had been brightly coloured, presumably to aid in rescue, but the ravages of leaking stove oil and time had rendered it a dull red. The effect was to make me feel like an industrial Boy Scout—ready for any contingency from getting lost to a rare bird sighting to the happenstance discovery of a major ore body.

The broken rock smelled of mashed elements—vaguely like a multivitamin. A few pieces too small to even rank as shards hit my pant leg. I swung the hammer again, this time aiming for a small ledge. A piece the size of a small orange slice broke off. I slid the loupe, slung from a cord around my neck, from its protective case and studied the freshly broken rock. Where it had been mangled by the hammer it was difficult to note any detail, but on the freshly split face it was a mass of near black angular faces mixed with what looked like opaque glass. I had the horrible feeling that I should be fascinated by the micro-relations in the rock but, to be truthful, I wasn't. It was dull and industrial.

Awmack and I moved on, our own minor agents of geological change, splitting rocks, kicking them down slope. After just a few minutes I was puzzled why there were any mountains at all. All around us rocks, big and small, were falling down the mountain. Some were dislodged by us, but most gave way to some unseen force. The broken-glass crackle of sliding rock was with us constantly. Erosion is often expressed as a slow geological phenomenon but what I saw made me wonder why the valley below wasn't largely filled in.

A fellow could hurt his eyes up on the mountain, focusing on the microscopic details of rock then looking up to see a sweep of mountain on the valley opposite. Awmack held out a piece. "See those round dots, like fly shit? That's neotosite. It's real handy because you see it quite often when you don't see other copper minerals."

While we worked we could hear Sullivan clanging on rocks nearby. He glassed a chunk, tested it with his magnet, and then decided to take a sample. When the sample was bagged he wrote in the little yellow book that geologists and prospectors carry. He saw us looking and said, "How do you spell 'sericite'?"

At noon Awmack stopped under an outcrop and dropped his pack. Lunchtime. Sandwich in hand, looking up, he stabbed a finger in the direction of an outcrop. It was finely layered, like pages in a book. That indicated it formed under water. This was a good sign. If the sea floor stayed half intact, during the millions of years of tectonic upheavals, up ahead of us on the mountain would be shallower rocks, and ones more likely to be mineralized. Overall, he said, the property had the right conditions. "It has the plumbing," by which he meant it had the inner mechanism, faults and fractures by which minerals could have entered the host rock.

For lunch I had cucumber, havarti cheese. Chocolate bar. Apples. The bear bell in my pack had stabbed my apple. Awmack laid his lunch out beside him so that whatever he wanted was at hand. One of the minor skills geologists develop is the ability to find comfort in what appear to be very uncomfortable places. Awmack can flop down on an outcrop and stay motionless for thirty minutes. I moved and fretted like a child on a hard pew. To say that the view was in excess of 180 degrees does not do service to the vertical extent. At our feet was a long slope and below it the green valley and beyond that a rise of another mountain. It seemed incredibly remote. Awmack was long past the no-one-has-ever-been-here-before mind trip but I wasn't. As a geology student working on a property out of Atlin years ago, he climbed a ridge to have lunch with a top-of-the-world view. "I tripped out on being the first one there. I looked down. There was orange peel." He suspects that the most virgin ground he's worked is in Panama, where no geology has been done for thirty to forty years, if ever.

As we ate, Awmack looked across King Creek to the bluff where he'd been working with Avie. Below the campsite was a triangular gossan that they had explored, and below it a richer-looking gossan. All this was included in the claim Awmack had registered with the BC mineral titles branch under the name Adam. His company had claimed so many units; each unit is five hundred metres by five hundred metres square. In return for the mineral rights to the area, Awmack's company has to do the equivalent of one hundred dollars per unit in exploration work. Otherwise it reverts to the Crown and someone else can claim it. All research is publicly available, so knowledge of the property builds.

Peering across the valley, Awmack let his thoughts surge ahead. How, if ore body could be proved up, there would be too much overburden to remove economically. I asked about the history of the property and he said that the best exploration work had been done in 1975. But the company had two big problems. One was that the porphyry deposits were typically spherical—baseball-shaped was the phrase used—and that whatever the shape of the ore body here, it wasn't spherical. Another was that the Adam property was then eighty kilometres from the nearest road. Building a road eighty kilometres through this kind of countryside was prohibitively expensive. With the development of the Eskay Creek mine, however, a road head was twenty kilometres away. But, he said, geologists had to be cautious about spooking themselves from distant properties. The mining giant Freeport–McMoRan had a property for fifteen years in Indonesia and didn't drill it because, at fourteen thousand feet and deep in the jungle, it was inaccessible. Accepted wisdom at the time was that normal grade porphyry deposits would never be economic to mine. A new manager took over and opted to spend millions on a drill program. The mine that developed is the highest-grade copper-gold porphyry mine in the world.

The thing that got Awmack excited over the Adam property is its similarity to the nearby Kerr property, which is even farther from road access that the Adam. The Kerr, too, is hosted in monzonite. He got up from lunch, stretched, and banged off a piece of rock. "Andesite," he said. "Now I'm confused."

Later that afternoon Awmack took a ranging tour around the

talus slope on the north side. He came back saying he had changed his mind, that the dacite was andesite. "I'm the type of guy who can look at something and think, 'I know what it is' then look at it some more. Here, I decided the emperor has no clothes."

At 10 a.m. on Thursday, Awmack found some siliceous argillite—soft, flaky like a pie crust. It had what he called "the qualities of Eskay Creek" about it. The mudstone meant we were in a marine environment. Awmack could see the transition from andesite. We were moving into a basin setting. He wanted subaqueous volcanics. Think Guam, he said. Volcanics out of a deep ocean trench. Sulphides dissolved in hot volcanic liquid spewing into the cold sea water, slowly forming mineral pipes. The pipes reach unstable heights, fall over, and new pipes form. "You want the rock to have been formed in two hundred to a thousand metres of water," he said. A general rule in volcanics: if it is red it was formed in atmosphere (red is hematite). If it is greenish the rocks were formed under water, where chlorite has coloured them. He looks at the rocks nearby. "This range we are in has both. It is ideal."

What Awmack was doing—pondering the dacites and andesites—is known as basic geology. More important to the project, however, is how these observations fit into the overall understanding of how mineral deposits are formed. Every deposit forms to a rare combination of a dozen or more chemical, structural, and tectonic factors all coinciding in one spot. What's more, each deposit is different from any other. However, there are families of deposits that share key characteristics—they formed through the same basic processes, just manifesting a little differently through their unique conditions (for example, what if metal heavy fluids pass through carbon-rich muds instead of limestones?). There are probably two hundred porphyry deposits along the Cordillera from Alaska to Chile; while all differ slightly they also share common features. Every deposit that is discovered stretches the range of things to watch for. Every geologist considers some features of prime importance and some as incidentals—think Awmack pondering the dacite.

The nearby Eskay mine gives us a good picture of the thought process of trying to visualize deposit formation and deciding what is important and what is not. Its genesis was simple. There was an undersea volcano. Hot water boiled through and around the rock. Hot water can hold lots of metals in solution, including gold, silver, lead, zinc, copper and iron. The metal-rich fluids got channelled into faults in the rock and rose up toward the sea floor. When they got to the sea floor they vented as black smokers—narrow towers emitting hot water. Under some conditions the metals in the water can dissipate. Or they can form little chimneys (this is happening off Vancouver Island today). If vented at depths of less than one thousand metres, the ocean pressure on the sea floor is low enough that the hot fluids boil. Boiling is one of the main ways to deposit gold out of fluids, so the gold precipitates.

We made our noisy way across talus. On the other side was a messy outcrop that Awmack declared as chert. The mountain here was minimalist, elemental. Tiny alpine flowers grew in crags. A patch of dirty snow stuck like a scab to the north side of the outcrop. Everywhere water was moving: in trickles off the rock, in rivulets and streams. Across the valley water was tumbling off the mountain like it was having a shower. The mountain was enveloped in dreadlocks of water, braiding the mountain. Between the streams there was green, equidistant from the streams.

After lunch I busied myself with a contact between argillite and tuff. I found altered rock, and dug between the two—thinking what Tom Bell said about a badger. I flipped my rock hammer over and used the chisel end to gouge out the contact. According to Awmack, previous geochemical surveys had shown that there was anomalous gold. So the gold had to be coming from somewhere, didn't it? I found a piece with luster. Too much for chalcopyrite. I took it to Awmack. "So you found the ore body, did you?" he said. He glassed it for a moment and handed it back. "Lichen," he said.

I wandered off on my own, knowing that as long as I did not go down the mountain I could not get lost. We were, in effect, on an alpine island. I worked around one bluff, then across a talus slope, stirring ptarmigan. Fog had moved in and distorted the ever-weakening sound of Awmack's rock hammer. I saw a bluff that turned out to be a rock ledge three paces away. Perspective of all sorts can disappear in a camp. One geologist, having worked long and hard in the bush, looked up to see a grizzly bearing down on him. He shrieked, then realized it was only a porcupine. Pat Suratt told me about a prospector who was accompanied by his dog, Rommel. When Rommel was disciplined by the camp's drillers for eating their steak dinners, the prospector took a chainsaw and hacked the tent to pieces. Another old prospector was found in a tree, handing out imaginary three-piece Kentucky Fried Chicken dinners. And the crew of yet another camp knew it was time to fly out one of their staff when he reported, in all seriousness, that his rock hammer had the ability to sing.

Across the mountain from us, so deep in a draw that he was out of radio contact, Tim Sullivan pulled his yellow notebook from the left-hand pocket of his geologist's vest and wrote in it the following: "Sample #77485. August 28/02 Adam North. Host Rock: Felcic Volcanic poss. Alt. Monizinite." Then, under comments: "lots of pyroitite very different rock than what I've seen on the south end just above camp. GPS 051. Elev. 1200."

Then he put his pen and notepad away, dropped the sample contained in a marked plastic bag with the others in his pack, shouldered it, and stepped on his way. In the course of a working day he might make fifteen to twenty such entries in his book, and take slightly fewer samples. At the end of the season, when Awmack writes his report, he'll use Sullivan's data. In Sullivan's notebook, there is a sense of the rock, and the man that made the notes:

- "Very fractured rock right against the dyke. When you dig at it it's very rotten and a bed of malachite appears mineral has gone."

- "Nice serricite Alertation with deceminateed calco. look like silver is present. poss. VG can't see well weather lousy."

- "Host rock: Fucked up Bleached Out Monzonite"

Short, muscular, and brimming with attitude, Sullivan looked like a modern Marlon Brando filtered through a junior hockey change room. He wore a baseball hat backwards with two ptarmigan feathers stuck in it. He was thirty-two. He drove the biggest, baddest truck in all of Hazelton, a blue 1980 Chev short box, three-quarter ton. In the winter he pulled logs out of the bush with it and cut them into firewood. In the spring he went to Vancouver with his girlfriend and sat for hours in Commercial Drive coffee shops, inhaling the urban culture. "I used to be sympathetic to the anti-mining arguments. That's when I wore my Che T-shirt. Those were my feet-on-the-neck-of-the-oppressor days." Several things happened to change his mind, the most important of which was the discovery that he had a remarkable talent for outdoor activities. Although he now has certificates in outdoor guiding, there was a time when he was a ratty kid in Hazelton. His father, a legendary wrestler, sent young Tim to an Outward Bound camp. The camp finished with a required two-day solo. Students were forbidden to eat or swim. Soon after his instructors left him, in a pup tent beside a lake, Sullivan swam out to a nearby rock. He watched as two hikers passed within fifteen feet. "I was into the ninja thing and made myself invisible," he said. "They walked right by me. They never saw me. I went: 'All right! It works!'"

He grinned a lot and used pop culture words when discussing the mineral exploration industry; of the downturn in mineral prices he said "Earth to deathstar." In BC mineral exploration circles, Sullivan had certain renown: as a weird guy, for great strength, for being one of the best young prospectors in the province.

"As soon as I see a person I'm reading them," Sullivan said to me one afternoon. "I really have an eye for details. It is one of the senses I'm blessed with, this talent for noticing things, especially odd things." From the densest bush he recovered muskets, stashes of flour, an old bullet-lighter. If there is something odd about a person or place, he notices it. On the Adam he's found mountain goat wool, which he's stuffed in his pocket. He dreams of wearing a sweater made entirely of mountain goat wool.

One afternoon Tim and I were headed down a talus slope that ended, disturbingly, in a cliff. Below the cliff was the white, rock-strewn frenzy of Terwilligen Creek. Sullivan fearlessly bounded to the edge of the cliff then patiently watched as I edged up to it. I apologized. "You've got to be crazy not to be scared, and if you're not scared I don't want to fucking hang around with you," he said. On the ring finger on his left hand was a ring with a fish with waves around it and a kayak. He said it represents his interest in the outdoors. He's rafted almost all the major rivers in northern BC.

He's strong and knows it, but he'd never tell you he was swift of foot. One time he thought he'd like to enter a triathlon. He was training and in a race. There were just two of them. "Look at me, this isn't a running body. I was running behind a long-legged runner, and the runner said, 'What's that noise?' It was my feet. I run like a duck."

We reached the rocky banks of Terwilligen Creek and turned downstream. The creek was an insanity of white spray tumbling through boulders and crushed rock. Ahead of us was an aspen copse. Sullivan called, "Hello Mister Bear! Hello!" He's respectful of bears but not afraid of them. If they know you are in the area, he says, they'll get out of the way.

After just a few minutes of pleasant flat-ground walking we turned to climb back up the mountain. It was steep enough that if you slipped you would in all likelihood die. Sullivan ascended as if on a rope—driving the tip of his rock hammer into a crack here, clinging to bushes, finding toeholds in the slightest crack. Again we paused. Across from us a glacier was leaking a white stream from a dirty thin edge. That edge, said Sullivan, was a sign the glacier was receding. That meant new ground, never explored. He wanted to be there.

Farther up the slope he found an outcrop that interested him. He studied it with his loupe and said, "Sphalerite. The deceiver. I'm lucky. I have a good eye for it. It mimics whatever mineral it is near. In this case it is mimicking hematite."

Sullivan is very competitive. As a soil sampler, he used to run, insisting on getting more samples than anyone else. When he found an interesting rock, he brought it back to camp. When, at the end of a long season a friend complained about boredom, Sullivan said, "Dude, aren't you interested in the rocks?"

A week after I arrived at the Adam property we reversed the set-up process, bundled the tents and hundreds of pounds of gear into net slings, and demobed—which is jargon for demobilized—back to the Eskay road. Awmack had to analyze samples and study his notes, but his hunch was that the property did not have the combination of access and geology to convince investors to sink millions into an exploration program (he was right).

I drove off, bound north for Dease Lake, to see Jim Reed. Reed lives in a honey-coloured log cabin in a little oasis of civility at the end of the utilitarian airport. Jim Reed is a helicopter pilot, and one of the most respected in the business. Every northern town has a Jim Reed. They are like raptors. Each has his own area, his own zone. When they get the call they soar. He was able, I was told, to set an airlifted fridge down on a platform within inches of its target. A geologist's assistant told me about a time when Reed toed his chopper into a particularly difficult slope. The slope was so steep the blades were cutting the grass on the outcrop in front of the machine. The assistant and the geologist clambered in and belted up, and Reed spun the helicopter deftly away. All they heard in their headsets was Reed's rich Kiwi accent: "Fishin's good around here, oi?"

Reed was in his fifties, silver haired, suntanned, reserved but not shy. When I met him he was having a dinner of chicken, rice and broccoli after a long day. Hanging from the log beams and against the walls of his log home were fishing rods. It is fair to say that a helicopter pilot in northern BC accesses the best fishing spots in the world. It was 9:30 p.m. He had started the day by moving two Discovery Channel photographers off volcanic Mount Edziza. Then he met a float plane at Forfer Lake and slung in six external loads and three internal loads of geological exploration gear. At any time he may get air ambulance work out of Dease Lake. Recently, he flew out a resident hunter who had broken his leg; two days later he flew out the body of a Texan who had a heart attack while hunting near Todagin Mountain. The Texan's body had slipped down a mountain, so Reed flew in an RCMP officer who set a line on it and lifted it to a place where he could slip it into a body bag. Reed usually goes to sleep at 9:30, and is in the air by 6:30 the next morning. He flies 130–140 hours a month in the summer, 600–700 hours per year.

"A lot of pilots won't fly here because of the mountains. I think the mountains are a delightful place to fly. The trick in mountain flying is knowing where to expect turbulence, where to expect updrafts," he said. "Don't fly down the lee side of a sheer mountain, because there will be a downdraft. You stay to the windward side, where there is more lift." The pilot's challenge is to read each valley, each bluff, with a mind to what the local manifestation of the wind will be. The prevailing winds are westerly, but if they shift to northerly then everything changes. A face that may have been washed with uplift can be curtained with a treacherous downdraft. Fog is a menace. Fixed-wing pilots envy helicopter pilots for their ability to set down quickly in fog, but as a rule fixed-wing pilots don't operate fifty feet from a cliff face. Reed said that the trouble with fog is that you need to go slow. When a helicopter goes slow it doesn't fly so well. Mountain passes are treacherous; they may be clear on one side but not on the other. His answer: always have an escape route and extra power.

Reed has had one serious accident. He was working in New Zealand, lining cut brush that had gone to seed. Airlifted to a barren railway embankment, the brush would reseed and help stop erosion. The trick in such work, he said, was to rise fast, so the line would cinch on the bush. Somehow the line bundling the bush got slung over his skid when he set down; when he soared aloft, the weight of the bush torqued the helicopter upside down. Reed landed in a swamp, which may have saved his life. He hit his head on the control panel and it half-scalped him. That was fifteen years ago. A shiny half-moon scar marked his forehead.

Reed and his wife, Sharon, both have rolling New Zealand accents. He grew up on a dairy farm in a small town near Hamilton. His father was a private pilot and sometimes Reed flew with him. He got a job as a deer culler, working in remote camps where supplies were parachuted in bags stuffed with straw. It was there that he got his first ride in a chopper. "That was it," he says. He gave up on dairy farming and concentrated on commercial flying. He flew geologists up Mount Ruapehu and throughout Tongariro National Park. Then he took on commercial flying, spraying fertilizer on cabbage and broccoli, herbicides and fungicidal sprays on pine forests.

Reed has been flying the Canadian north since 1990. He started at Burns Lake and also worked in Fort St. James. When his employer, Pacific Western, bought out Yukon Airways they asked him to run the Dease Lake base. He's flown for all the big companies and most of the small ones too. When I met him, he had twelve thousand to thirteen thousand hours, but it may have been more. "I haven't added them up for awhile," he said.

If you want to see him levitate without aid of a helicopter, ask him about tourists. He said tourists want to walk into tail rotors. After a flight, and despite repeatedly being asked not to do so, they slam seatbelts in the doors when they get out. In the air the metal clasps do the eggbeater on the fragile bubble. River rafting jobs include many tourists with the added complication that the rafting guides seem determined to overload the cargo. It has got so bad that Reed will not even give them the big net anymore, the one used for transporting empty fifty-gallon fuel drums. A lot of people don't understand helicopters, he said. The trick is shifting to translation. When a chopper hovers it is pulling air in, but when it goes ahead at six to eight miles per hour, the air is coming to the rotor. "A lot of people have this idea that helicopters can go straight up; they can, but not loaded."

The way a lot of younger or less experienced pilots talk, you get the sense that you are beside them in the seat and that the machine and what they do with it is central to the story. But Reed has been flying for so long that he and the machine have vanished, and all that is left is the point of view, over the wilds. Scenes: of a grizzly sow suckling two cubs in a snowbank, a moose islanded in a pond and surrounded by five snapping wolves, osprey drilling into a lake. The telegraph line, old cabins, the workings of the Chinese miners around Thibert and Goldpan Creeks. Their rock walls are still standing.

The next morning I drove north from the hotel in Dease Lake for several blocks until I came across a restaurant. Three big rigs, outfitted with substantial-looking bear and moose guards, rattled in the parking lot while their drivers slept. I paid for coffee and asked the only customer, an enormous young man, where I could fuel up. "You're out of luck," he said. There was nothing open on Sunday. Unless I could bum gas bootleg, I'd be staying in Dease Lake. I walked out of the restaurant, stood at the side of the empty road, and looked south. There, just beyond the hotel sign, was a service station, "Open" sign swinging in the breeze.

On the way back to Smithers I saw four black bears.

Lorne Warren lives in a two-storey white and brown house that overlooks green fields and a bend in the Bulkley River. When I went to see him—mid-afternoon of a warm grey day—he was standing outside, with one hand thrust in the pocket of his dress trousers and the other clenching a cigarette. He was supervising two young men who were hiking goods out of a garage. As I got out of my truck one of the young men said, in a voice big enough to be heard in Hazelton, "Do you have any idea where the blasting caps are?" Warren looked at me and my notepad. He didn't look happy. I suspect that blasting caps are supposed to be stored under other conditions.

Earlier in the year Warren had won a prospector's prize, and a *Maclean's* magazine writer had had him on the phone for three hours. From that interview came a quote in a small story about Lorne taking the clothes out of a chest as a child so he could keep his rock collection. He was wondering if our time was to be similarly wasted.

I looked out at the fields and said that this was the way to farm: have someone else do it while you looked at the fields. He said that as a kid, growing up on a ramshackle farm on the ridge behind us, he'd had to clean and candle eggs from a flock of eight hundred chickens. For these he got three cents a dozen. He has had a deep hatred of chickens ever since. And a belief that there had to be a better way of making money.

We walked into his basement office, which was stacked with papers and rocks. At one time, he said, he had fifteen to twenty mineral properties. He had two thousand units. But NDP policies and rising claim rates had made such ownership prohibitive. So he has reduced to two hundred units. That still costs him forty thousand dollars a year to maintain. He's hoping the markets will improve and has reduced his inventory to key properties.

When Warren was a kid, Smithers was still a mud and track town. On Sundays, he and his brothers hunted for change in the open ditch that ran in front of the Bulkley Hotel, where fighting drunks had spilled coins the night before. His parents ran the Corner Café, a small-town coffee and burger joint.

These days, there is probably no one who knows more about the Smithers region than Lorne Warren. In a government office in Smithers, he found volumes of Ministry of Mines books. What he learned from these books often dovetailed with what he had seen on the ground.

An old-timer named Art Cope had a cabin and claim on Dome Mountain. He asked Warren to walk in and put mothballs in the cabin. Warren hiked in on a Sunday. Looking around, he saw that someone had been working the claims, high-grading. In an old apple box he found quartz. Warren banged at one. It was hinged—held together by free gold. Another old-timer had shown Warren how to run his fingers over the sample. If there is native gold, silver or copper present, it will have barbs. Warren calls them hackney. The crystal structure had barbs. Warren took a sample, crushed it and panned out and half-filled an Alka-Seltzer bottle. Much of the area was tied up in Crown grants, but these were let go in the 1970s. When the price of gold went up, the Crown grants were put up for auction. Warren bought up Free Miner licences for eleven family and friends. When the names were drawn out of a hat Warren or his cohorts had six of thirty-one Crown grants. He optioned the properties to Reako Explorations.

Warren tells me that he has thirteen to fourteen properties optioned. Last year he had two. He's had five days off since April. There is a sense of optimism in the business, an odour that a man like Warren, with his time-honed sense of direction, can raise his nose to. A total of $1.4 million is being spent on his properties. The biggest project, for $350,000, is being paid for by Rubicon Minerals, which is drilling the Axelgold property. Another property that he has hopes for is the Foremore, which he picked up when Cominco got tired.

He conducts his business from an office overlooking the same fields he used to stare at as a kid. He grew up in a house on a ridge above us. He got interested in prospecting via a neighbour who was a packer—he carried flour and supplies into a camp on mules and packed the ore out. In a shed in the back of this neighbour's house he kept ore samples. "Kids like shiny things," said Warren. He liked them so much that he went home and, as the *Maclean's* article reported, he pitched his clothes out of a drawer to make room for his rock collection. On Sundays, the only day he had off from the farm, Warren poked at fossils in the Driftwood fossil beds or went on tours of the local mines. He assembled an idiosyncratic but unique knowledge of the area.

Among the neighbours who lived near Warren when he was growing up was an old-timer named Pete Berg. Berg claimed to have a knack for dowsing. Pete was old—how old nobody knew, but maybe eighty-five or ninety. Warren walked to Pete's place after school and played crib. Warren had seen Pete dowse for water and tried it himself. But he didn't have the touch. So Pete took Warren's hand and thereafter Warren could dowse. Something got passed on, like the culture for sourdough bread. When Pete wanted to dig a well, he took from his trousers a gold watch with a chain and let it swing. It swung twenty-seven times. They dug, first through topsoil, then into clay. Down in the hole, Warren scraped it into bags. "At twenty-six feet the well was bone dry, at twenty-seven feet we hit water."

Warren's father died at the age of forty-eight, when Warren was fifteen. His father had never been healthy—years in a Polish POW camp had reduced his weight to ninety pounds. Lorne was the oldest of eight children. The youngest was eight months when his father died. All the family had to live on was a two-hundred-dollar-a-month pension and it came with a hitch. If a neighbour helped out or someone made a donation, the amount was deducted from the pension. To help out, Lorne took a job with Art Cope, the old packer. He took a prospecting course by correspondence. He and Art optioned a property in the Driftwood Mountains. They called it the Rainbow Claim and they convinced an outfit that it had promise. They split a five-thousand-dollar option payment. Lorne gave some of the money to his mom, bought her a dryer, invested in an outfit called Native Mines and doubled his money. That same season he took another job with Gordon Hiltchie, on Dome Mountain. "We flew in a small helicopter and stayed until the bug dope ran out." They actually ran out of bug dope before they left.

They would swat a bare hand and kill a hundred mosquitoes. They were camped in a swamp. They walked eighteen miles to Telkwa. This was 1963.

Warren became so involved in the mining business that his principal at school made a special allowance for him: he could leave school earlier in the spring and come back later in the fall. When Phelps-Dodge showed up in the Stikine as part of the copper porphyry rush, Warren took a job with geologist Tony L'Orsa. L'Orsa was a highly regarded geologist with an unusual handicap: he was colour-blind. But he quickly discovered that Warren could be his seeing-eye dog. Warren was hired as camp cook, so he'd get up early, prepare breakfast, then catch the last chopper out to prospect. When he found a couple of showings they made him a prospector instead of cook.

During the 1970s Warren made a good living finding properties and shopping them around. "I was called the McDonald's hamburger stand for properties." At one point in the 1980s he had twenty-five properties optioned. The average option fee is five to fifteen thousand dollars. Some prospectors want thirty thousand but that's too much. The most he's ever received is twenty thousand. Like a lot of prospectors, he's taken shares in lieu of cash payment. Warren's strategy is to take a down payment then buy stock in the company. When the company makes an announcement about the optioning of the property, the share price rises. "It's almost insider trading, but it's not," he says. While the stock market was booming Warren did well. Maintaining an inventory of properties cost him a quarter of a million dollars a year. When the market collapsed, he was left with some big bills. "I had a big-company mentality on a prospector's budget," he confesses.

I asked about the dangers of prospecting and he told me

about a time he and his son Chris were staking the Bill property, 290 kilometres north of Smithers. They wanted to put a post on a cornice about 1,500 feet above the valley floor. It was winter. There was a lot of snow. The pilot swung the chopper over the cornice, and Chris opened the door of the chopper and tossed out a claim post and a length of rebar. Then he stepped onto the landing bars of the helicopter and onto the snow—and vanished. Warren peered across the seat. There was a hole three feet across. With a sick feeling, he had the pilot circle away and under the cornice. If there was a hole under the cornice, then he would know Chris had plummeted through and would be dead on the cliffs below. They searched under the cornice, saw nothing unusual, and rose again. Warren took a rope and tossed one end down the hole. It was twelve feet deep. Chris attached himself and they pulled him out. He had sore knees, but he had staked the claim. He later enrolled at Red Deer College; he didn't seem interested in prospecting. Warren hoped he would change his mind: "He knows more than he thinks he knows."

Ever since he was a kid, Lorne Warren has had an eye to what's around the next bend. His bedside table was stacked with Arthur C. Clarke, Robert Heinlein, Isaac Asimov—futurists. So, in 1978, when gold was in the middle of a long ascent in value, Warren went out in his backyard and built a rocker—a gold-rush-era device for separating gold from sand. He knew that if gold was to continue to rise in value that there would be a rush for placer gold. A year later, when gold hit seven hundred dollars, he hired a helicopter for ten days. He had in mind the area about forty miles west of Takla Landing. Warren had worked in the area in 1971 and knew it held promise. His goal was twofold: to stake as many claims as possible and to identify a claim that he could work himself. At Kenny Creek he walked up the stream bed and came across a hippie family. Back-to-the-landers, they were brewing dandelion roots for coffee. They had built a rudimentary sluice and were digging holes in an old access road. Warren walked over to their sluice. Lodged in the third riffle from the bottom was a half-ounce nugget. "They said, 'One claim will do us a lifetime.'" Warren staked the surrounding area. "A lot of people were pissed off when they came in the spring because I was already there," he said.

For several years Warren devoted his efforts to placer mining. Because he was already established, he was able to observe a rush from the unusual perspective of someone already there. In the first year 120 people visited the camp. So many people got their vehicles stuck that Lorne put up a sign. "We work from 8 a.m. to 5 p.m. We won't pull trucks out of the creek until after work." They wouldn't take cash for salvage, but a bottle of scotch or a newspaper not more than two weeks out of date was acceptable. One day after high water passed, Warren opened the pantry in the plywood shack. There were fourteen unopened bottles of scotch. They threw a party and thirteen people showed up.

It was good work but hard work. Out of 120 operating days, he worked 53 days. Much of the rest of the time was spent under the crawler. In the first year he retrieved one hundred ounces. That which was in nugget form he sold for as high as nine hundred dollars an ounce. The gold dust he sold to Delta Refining in Vancouver. The operation was running smoothly but something told Lorne to get out. He found a buyer in Williams Lake and sent his last shipment, worth $60,000, to Delta Refining. When the payment was delayed he phoned and begged for the money. The cheque they issued was likely the last. Two weeks later they were broke.

When Warren thinks of future mining projects he has in mind BC Hydro's program whereby a mine can develop its own power from a river, or even sell surplus power. There are more than a few deposits that he knows of that hinge on the availability of cheap power. And he says he's read of a new process for producing copper on-site. Those are big-picture items. To survive in hard times, he's developed a log house recaulking and refinishing process. It keeps the cash flow and keeps his prospectors hired. Tom Bell has worked for him. Recently they were working on the log house of a Vancouver developer when one of Warren's employees stepped on something hard in the flowerbed. It was a piece of native copper, very rare. I asked if he knew where it came from. He said, "Not yet. But we are working on it."

Howard Bradley knows a lot about open-pit mining and a bit about tortoises. The manager of the Huckleberry copper-molybdenum mine outside Houston, Bradley used to run a mine in California that encroached on the grounds of an endangered tortoise. Bradley and his staff took a course in turtle ecology and learned how to handle them. And it worked, too, because during Bradley's term there only one tortoise died, and it was from causes later determined to be natural.

I'd come to see Bradley because I thought it would be useful to see a mine—the goal, after all, of exploration efforts—in action. How do you go about extracting minerals from the ground? At what point do the valuable rocks get separated from the junky rocks? What does copper look like before it is a pipe under your bathroom sink anyway?

Bradley told the story about the turtles in response to a question about the environment. At the time he was standing in the corner of his mine site office, in a steel frame building that ever so gently telegraphed the vibrations of the mill next door. He's a Wisconsiner by birth, and a practical dresser by habit—worn jeans with a solid, unbraggy belt, green work shirt—so the overall effect is of take it or leave it. If he had been wearing a tie, I might have thought he was telling us the story about turtles to impress me as I wrote my notes. But I don't think so. Who knows?

Three of us were in the room, having just signed in. Dave Caulfield—geologist, principal in several Vancouver exploration outfits and, in his capacity on this day, director of the BC & Yukon Chamber of Mines (BCYCM)—was giving a tour of the north to Dan Jepsen—the new executive director of the BCYCM, and the man charged with levering the industry out of the public relations doldrums.

Bradley's office was the size of a corner convenience store, bare but for a white writing board wherein the three-dimensional goings-on out his window were rendered into two-dimensional flow charts. He was about to tell us what we were going to see before we actually saw it. His window viewed onto a grey rock hillside from which a chunk had been removed. It looked like an apple with one bite taken. At the bottom of the pit, excavators metronomically swung bucket after bucket of rock into waiting trucks. Twenty-one thousand tons of rock run through the mill each day. That's a modest amount by modern milling standards. There was nothing for scale reference, but looking out the window you just knew the equipment was outsized.

Bradley stabbed a red felt marker against the white marker board. Before digging the rock from the ground, the miners have to decide whether it is ore or just rock. To do this they drill holes in a pattern, nine metres apart, twelve metres deep. A sample is taken from each hole and assayed for copper and molybdenum. If minerals are present in the right amounts, the rock gets run through the mill. If not, it is stripped and trucked away. From this elemental beginning, the mine process splits in two. One has to do with the milling of the ore; the other, and more controversial, has to do with the manner in which the waste rock is stored. He explains this to us; then, because he is too smart to ask if we understand, says "Okay?" We all nod. "I'll get my boots," he says.

The first of many levels of weirdness of an open-pit mine is the fact that a lot of machinery is electric. Diesel generators are connected to drills and excavators via massive black umbilicals. This means that, if you really wanted, you could pull a trick on the night shift and unplug the equipment. Dan and I were both taken with this fact, and while Howard and Dave looked at the drill we went over and stood by the cord. We were both cautious, as if by touching it we'd get a shock. After awhile we decided that since nothing was on we could touch it. It was the size of a man's thigh.

The area we were in was originally intended to be a waste site but in one of the flukes that often happen in exploration, drillers had discovered that one of the areas slated for a disposal site was actually rich in ore. The rock here was shattered—both as a result of the blasting and as a result of the ancient trauma that had created the ore body in the first place. The plan for the mine had been rejigged.

From the blasting site we drove—carefully, because the truck radio was not working—through the yard. By industry standards the trucks at Huckleberry were not large, but they could easily have driven over us. You have to think that somewhere in the working-class annals of open-pit mining, there is an account of a fired employee who vented his feelings by driving an ore truck over the boss's pickup. I know that in some mines the truck drivers read paperbacks while they drive, so slowly do the trucks move. We paused at the intersection of two gravel roads to wait for a truck to lumber past. I peered up. It was big enough to have a porch, like a cabin. A pretty woman smiled back at me.

Of the tour that followed at the Huckleberry mill and the details of the process, I cannot honestly recount. The tour was conducted by an earnest engineer named Wayne Fong, and between his accent and the 100-decibel-plus roar of the ball rollers, I picked up three words: "digester," "instant death" and "molybdenum." So what follows is based on a conversation I had later with Wayne, and my impressions of the mill.

After the trucks tip their load into the mill, it is sent through a grinder and shot along a conveyor to several ball grinders. Looking like giant rock polishers, these further reduce the rock to powder. Though I am no engineer, the tonnage and rotation speed of the mills made me dwell on the physics of bearings and how, should the ball mill ever even slightly wander from true, it would tear itself loose from its moorings and take off like a spring-loaded toy. From the ball rollers the rock powder is mixed with water and passed through a series of flotation cells. Fong's role here is that of the brewmaster, seeing that the mix is right. Of special importance is the viscosity of the slurry, for the separation of the minerals is contingent on bubbles catching the material. The mix got more and more attractive, looked like a St. Patrick's Day special ale brewed by a microbrewery. The digesting is done in a series of pots not dissimilar to hot tubs. The slurry at the top is caught in bubbles, and is carried into a trough, where it is passed to other pots in a series of refinements. The final stop in the mill itself was at a long mechanical-looking thing like an accordion, where liquid was squeezed from the mix and the concentrate (a mix of copper and gold) tumbled to a room below. The molybdenum is separated elsewhere. We trotted down steps to the warehouse, where we yanked the earplugs from our ears and stretched our jaws. After the shriek of the mill, a human voice was disproportional and wonky. From here, Howard explained, the mix would go to Japan. The pile was no bigger than what you might see of manure outside a farm. For the first time in the day I thought something was small. I asked if I could have some. I put it in my pocket.

To see what became of the tailings—the ground-up rock from the mill—we bounced down a long curving road, jolted by sharp rocks that slide under the tires, until a point about midway, and got out. This, said Howard, using the closest thing to a gesture I'd seen all day, is the dam. The dam is three kilometres long. "The concern, well, the issue, is ARD—acid rock drainage. All rock with the potential to generate acid will be under water. If oxygen can't get to it, it won't produce acid." He points to the bank opposite. Trees and vegetation have been stripped away and a line of orange markers planted along a topographical line. That will be the eventual height of the dam. Where we are will be buried in the dam itself. Someone asks how many years the dam will last. Howard says: "A long, long time."

Gunning a rented 4x4 down the mining road on the way back to Smithers, Caulfield summarized his complaints about environmentalism. Mines like Huckleberry and Eskay, he said, have an imprint on the landscape about the size of Richmond's

Lansdowne Mall, with one important difference. "The mall will never go back to bush; the mines will," said Caulfield. North American society cannot function without metal. "Nobody, not even David Suzuki, is interested in going back to the loincloth." We pass plantations of second-growth pine. Geologists are hesitant to put themselves in opposition to loggers, but Caulfield can't resist. Logging, he said, has a much larger environmental impact than mining. "The greatest value to the province per acre of disturbed land is from mining."

We caught up with a lumbering logging truck. We were eager to get back to Smithers and get out of our soggy boots. Jepsen, monitoring the traffic radio, said "Go!" Caulfield stepped on the gas. It seemed like we were going to impale ourselves on the overhanging logs.

"Watch out!" I hollered.

"Go! Go!" said Jepsen.

Caulfield gripped the wheel hard and hollered, "Will you both shut up?"

The next day I drove to Prince George, where I was to meet the owners of Falcon Drilling. In the strange subculture of mineral exploration Falcon Drilling is an anomaly. Instead of the rock music that plays when you phone most drilling companies, at Falcon they play CBC classical music. The offices on a side street in Prince George, while not plush, are not smeared with grease, as is the case in many drilling offices. And here, the usual lunchroom literature of pornography magazines is replaced by the upper reaches of the *New York Times* bestseller lists. Clocks on a wall show the time in Peru and Argentina, where Falcon has crews.

Falcon Drilling has twenty drills working in Argentina,

Papua New Guinea, Canada, Myanmar and Panama. While I was at the office on my way south from Smithers, they were preparing to ship a drill to Korea. The senior of the company, Gary Paulson, is a quick-thinking man in his early fifties of medium build who has forged a role as a renaissance man in the exploration industry. He takes *The Economist* magazine and a bevy of newspapers. He single-handedly supports Amazon.com. He said, "Drilling is like buying a new car. Customers want it now, not in six weeks. That's why drills need to be shipped by air. It takes six weeks to send a drill by ship."

Falcon Drilling is Gary Paulson, brother Grant and a young innovative engineer named Bruce Hiller. Gary and Grant grew up on a farm near Dekker Lake. If something broke they had to fix it. So when Gary got into construction, he wasn't surprised or offended when backhoe operators witched for waterlines. If something worked, great, and to hell with the science that said it shouldn't work.

Gary is famous in the exploration community as an aristocrat and a smooth dealer, ready to shake hands with good luck wherever he finds it. He's met the president of Panama three times. Recently, Gary told me, a couple of lawyer friends were involved in a drug-related case in which a female had been murdered. It would help their cause greatly if they could find her body. They had an idea where it was buried—near Burns Lake. Gary has a couple of cadaver dogs. He let them loose in the bush but they got put off by a half-decomposed moose. "So they were no use," he said. He took coat hangers and walked through the bush. "I thought she might have a cell phone on her, or a ring. Enough metal, anyway, to make the rods cross."

He padded through the bush, lawyers trailing behind. The rods crossed. They dug. They turned up a nodule of iron pyrite.

"The lawyers weren't interested in the body anymore, they wanted to look for metal."

Gary Paulson seemed so delighted with his own thoughts that he might be his own best friend. He asked if I'd like to stay, as long as I didn't mind tagging along to his son's football practice. I followed him out of the river bottom that Prince George is built on and up a switchback road that emerged from a scrub of forest onto a plain of small farms. With an SUV in every driveway, it was clear this was not the pogey-and-firewood kind of acreage for which Prince George was once famous. The UNBC campus was just across a ridge. It was clearly a tony part of the city.

Gary shares a big house with his wife, a home economics instructor. The house was like a *Good Housekeeping* prototype, with everything organized and put in containers. My room was cleaner than any hotel I stayed in on the trip, and each room had a timer on its lights.

Gary and I picked up our conversation at dinner. He said the company's origins were in construction. Jempland Construction is named after the province in Sweden that their father came from. They started with mining road construction. They worked for Cominco on the Aley property; they built an airstrip at the Snip property. Their big break, however, came at Eskay Creek, where they were in on the early stages of a world-class mine and a rush of gold-rush dimensions.

Despite their efforts the snow was besting them. It came down in Brobdingnagian volumes. When opening a door to go outside became a major task the crew called it quits. They were preparing to leave. The snow came down faster than a man could dig. In the cookhouse, senior Paul Paulson, Gary and Grant's father, heard the news and curled his lip. He hadn't come over from Sweden to hear the word quit. "He called them fucking wimps and went outside to start shovelling snow to the outhouse," says Gary. "The geological crew inside felt so sorry for him that they went outside and started shovelling." The old man's determination convinced them to stay. Even so, it took four men shovelling full-time to keep the walkways open.

The dimensions of what went on at Eskay were incredible. Everything had to be flown in. They went through thirty to forty snowmobiles a year. Come spring they'd find lost equipment, including snowmobiles. "Stuff with a ten-year life span lasted a couple years," says Grant. Everything was geared to keeping the drills turning. A hundred-person exploration camp eventually arose at Eskay Creek. Six drills worked around the clock. A lot of summer jobs don't produce three hundred feet per day.

Staging out of Bell II, they flew in a Bombardier BR 400 with six-foot-wide tracks. It took six trips to fly it in. The mechanics descended on the pieces. It was working in eight hours. They used it to build a road, packed the snow with skidoos. With so much snow on the ground it was difficult to read the terrain. A Cat operator drove his machine over a cliff. A Sikorsky 64 lifted it back onto the ridge and it went to work again. They built the camp below pioneer Iskut prospector Tom Mackay's original adits. The first camp burned and they rebuilt. Come summer, activity increased to a frenzied pace. Choppers—205s, 206s, 500s. Ten drills ran around the clock. Promoter Murray Pezim flew in poster boy and former NFL star Mark Gastineau and nobody noticed.

Mining giant Cominco came to Falcon with a proposal: if Falcon could manufacture a lightweight drill capable of drilling to six hundred feet they could have the contract. The previous contractor had been using a Longyear 28, but it powered out.

The next model up, like the Longyear 38, weighed in at more than 1,200 pounds—too heavy for easy airlift. Such drills needed lots of power and special hydraulics. The trick was to come up with enough power to break the core off. A drill can have 6,000 pounds hanging from it. Also, it had to be made of parts that were easily available, as Gary says, "in Timbuktu." So they set about designing their own drill of readily available parts.

In many ways, Prince George is the Edison lab of British Columbia, the centre of the province's free enterprise inventiveness. The northern Silicon Valley. It was in Prince George in the 1960s and 1970s that the area's sawmillers pioneered small-log technology that enabled them to manufacture lumber at a rate that made the city a world power in two-by-fours. The residual infrastructure of that revolution exited all around the Falcon shop in the form of hydraulic and small-equipment manufacturers—in some cases a five-minute walk away.

The man picked for the job of redesigning the drill was a demure, self-effacing man of forty named Bruce Hiller. More comfortable in front of a lathe than in a crowded lunch room, Hiller is the son of a man involved in the Prince George renaissance. Trained in helicopter maintenance, he impressed the Paulsons, who offered him a partnership. He worked from plans in his head rather than any formal design, starting from the point that the drill needed to be as simple as possible. A driller facing a breakdown on the side of a Bolivian slope had enough problems without fretting over extra parts and tools. Systematically going through a drill's basic design he shed bolts wherever possible. Most crucial was the complex connection drive that he replaced with a much simpler universal joint. When the prototype—grandly named the Falcon 1000—was ready, they hauled it to a gravel pit on the city's outskirts. It worked, but sporadically.

Working under a tight time deadline, they faced a crucial test when a Cominco representative planned a visit to see the drill at work. Bruce and Grant set up camp beside the drill. At one point they set the surrounding logging slash ablaze. "It lit up the whole city," deadpanned Grant. After a month and a half of work, the drill was ready. On the days leading up to the inspection the drill ran well, then on the day itself it got cranky. Gary, delegated to escorting the Cominco man around, was put on delay mode. He talked and talked. They arrived in the pit. The Falcon 1000 was at work, satisfactorily roaring. When the drill reached seven hundred feet the Cominco man said "Enough."

Hiller's work paid off. The simplified drill proved to be a marvel to set up. Instead of a traditional half-day, it took a half-hour to set up. It can be set up as fast as a chopper can swing the parts in. At Snip, the Falcon 1000 chewed down to 1,200 to 1,400 feet. Never content with the standard, Hiller redesigned the next-model Falcon 1000, and the next. By the end of the year Falcon had six Falcon 1000s at work at Snip. They went in expecting a 15,000-foot contract, hoping for a 25,000. By the time they had finished they had drilled 100,000 feet.

A combination of unfriendly provincial policy and the after-effects of Bre-X have forced Falcon, like a lot of other drilling companies, to look elsewhere for work. "After the NDP came in we were advised to go overseas. We were supposed to go international," said Gary. "I asked myself: 'How the fuck do you do that?' Actually it is simpler than it seems." They tag along with an international company. Now Falcon has drills in Panama, Argentina and Myanmar. Said Gary: "I remember when Thunder Bay was an exciting destination."

The move was not without troubles. In Panama, for example, a driller's helper was kidnapped. Falcon crews had to take him to the hospital when he was returned. Falcon's policy is to not get involved, but this time they made an exception. "We explained we are looking for a mine. There's nothing to fight about yet." In a lawless quarter of Jamaica, Falcon crews needed protection from marauding bandits. They hired off-duty cops, the toughest of whom was legendary M16-toting Harry. Harry had shot seventeen people.

But nowhere is that work more challenging than in Papua New Guinea, a remote, jungle-infested area of incredible mineralization and challenging job situations. Gary had just returned from an inspection of the company's drill crews. A local war was being waged. "One side had a fifty-calibre machine gun and three bullets. The rest had bows and arrows. The warriors were creeping down along a creek bed. They fired one bullet and retreated."

"It is a thousand years behind in the evolution of man," says Gary, with typical straightforwardness. "They wear wraparound sunglasses. Sports jackets and bare feet. We bought boots for the driller's helpers—their feet are big, so they cut the sides out of the boots. Every Friday was payday—chaos, girls raped, stabbings. At one point we had to shut down a drill for a day and a half because there was a war on. The way they wage war is by yelling. So we had to shut down the drill so they could wage war."

Before we left for the football game he told me a story about working in Panama. A villager asked him for a $150 loan. When Gary paid the villager, the rest of his crew told him he was crazy and that he'd never see his money again. "But this Panamanian guy was really smart," said Gary. "Instead of buying food for his snot-nosed kids he had a portrait taken of his mother-in-law. Brilliant! The mother-in-law is happy. The wife is happy and then the kids are happy. It was brilliant! It was the best way to spend the money."

"Did he repay you?" I asked.

And Gary said, "No. Not really."

11 Roundup

The Mineral Exploration Roundup conference held in Vancouver each January is both the first page and the last of the annual report that western Canada's mineral exploration community writes to itself. For three days prospectors, geologists and financiers review the previous season's findings, study long-term industry trends and discuss future possibilities. It is much like any other conference in the trappings but for one important difference: rocks. Everywhere there are rocks: on display, in the complimentary suites (where the beer is packed in ice in the bathtub), core samples, grab samples, float . . . The conference is, one way or another, all about rocks.

I went to the show on a Tuesday, looking for a friend who had become a geologist. I walked into a large hall filled with booths, turned a corner, and was confronted with a table full of green cylindrical objects that even the most wizened old granny could identify. Dildos! Behind the arrayed goods was a large, muscular man with a beard and dark eyebrows and chocolate brown eyes. Harvey Waite is from Campbell River. He is a hardrock driller and sells core as a sideline. I looked at his business card, "U Wanna Rock." There is a lamp on Harvey's stall that is attracting a lot of attention because it is being given away. When someone makes a joke as they frequently do, Waite orders them to turn around. This is a bit shocking, given his size. He rubs the rounded core innocently along your back. It makes your legs wobble. The hardness is remarkable. Even the toughest knotted muscles give in to the rock. "The women laugh at it," says Waite, "you can hear them a hundred feet away." The massager is chlorite and epidote with some feldspar. While he talks he flicks pistachios in his mouth. He sells little rock pinnacles with toonies shellacked into the top.

Waite does not actually sell dildos but he does nothing to discourage the impression that his back massagers might have a sexual function as well. In the largely sexless trade show, the naughtiness around his booth is alluring.

Since 1972 Waite has drilled on oil rigs in Argentina and all over BC. For the last five years he's worked for Boliden, at Myra Falls on Vancouver Island. Maybe because the conference was crowded, maybe because drillers don't get to tell stories to writers very often, he said a lot of things, fast: like the time when he was working out of Stewart when the driller's helper stuck his hand in the drill. It was too late to fly, and the white of the bone glistened in the dark. At Bronson Creek on Red Mountain, he worked on a platform so high that if you slipped from it you would die. The terrain there was so fantastic that he remembers it over and above other places he has worked. A chopper that picked him up after shift filled with smoke. When the pilot didn't react, Waite said, as responsibly as he could, "Hmmm. Little smoky in here." The pilot looked at his gauges and said, "Maybe just an oil leak." Flying from the drill one day the helicopter got lost in the fog, so the pilot descended until the wash blew the clouds away from the rock. The pilot knew the mountain so well that all he needed was a glimpse. Once, the wind was so bad that he got stuck on the platform for three shifts. His helper went to fold a tarp and it splatted against his face, outlining every feature.

I wandered into the conference room and listened to a guy talk about a big copper deposit in Alaska. When it was over I walked back into the trade show but it was empty. The exhibitors were in another big room, swilling beer. The stands of many mining companies and associated suppliers were erected below banners and stalls. I picked up brochures and maps, helping myself, enjoying the ghost-town atmosphere. Tucked under the banner of the InfoMine table, sipping beer and munching a corned beef sandwich was an older man with white hair, blue eyes, casual jacket, green vest, beige trousers, white socks and white shoes. Overall, the impression was of a younger Bob Hope. He looked at me and said, "You better watch out: I'm radioactive."

Manny Mannex is eighty-three, and quite likely the oldest practising geologist in BC. Certainly he's in the best shape. Three times a week he attends karate classes at a school on Denman Street. He is a yellow belt. He reached into his pocket and pulled out a wad of dog-eared business cards and gave one to me. It was a picture of him, much younger. "I've been accused of looking like Einstein," he said. The picture strongly suggested Chicano chop shop. He held out a home-printed, scissor-cut business card with "Manny Consultants" typed in. "My father screamed at me. 'Why did you name your company Manny? Why? Your name is Emmanuel!' And I answered, 'Because ever since I was little people called me Manny!'"

He was born in Brooklyn on January 1, 1920. His life has been a series of cheerfully accepted setbacks. In the Second World War he trained to be a pilot, learning on Fairchild PT-19s and PT-23s, but someone in his class stole money and the whole class was washed out as punishment. So he became a bombardier. On the way to Hawaii in a factory-new Liberator his plane almost ran out of fuel; then, in Hawaii, another crew made off with the new plane and they had to settle for a used one. He flew many missions, on bombing runs over many Pacific Islands like Palmyra, Canton, Tawawa, Marshall and Gilbert. One of their assignments was to make repeated low passes over Japanese airfields so the Japanese fighters could not get airborne. His plane was attacked several times but Manny was never hurt. After the war he wanted to go into aeronautical engineering but his grades were not good enough so he went into geology at Columbia University. Specializing in uranium, he worked on magnetite properties in the Chicoutimi region of Quebec and in Newfoundland. On one trip home he discovered that his wife had mistaken his valuable collection of mid-1800s Indian Head pennies for a change jar. He moved west and hooked up with legendary promoter Murray Pezim. Pezim, he says, was interested in a property in South America. "He'd say, 'Manny! Go look at it!'"

As we talked people walking by said, "Hey Manny!"

Though he is retired his troubles are not over. At one point he owed Canadian federal taxes totalling one hundred thousand dollars. Using his OAP cheque he has paid that down to sixty-six thousand. The problem with the current mineral exploration, he says, is "No money!" Exploration needs money. He is mostly retired but has several projects, including one in Tonopah, Nevada, and the Margie claim, near Hope, which is one of the oldest claims in BC.

I liked Manny and enjoyed our time together. I like airplanes, and we talked more about planes and less about geology than we probably should have. When I got up to go I offered my hand. Manny shook it with a surprisingly firm grip and said, "I've got a tape recorder and I recorded our conversation."

The next morning I found my friend standing behind a booth. Mark Baknes was studying a paper on the Thorn deposit. Core samples from the Thorn were set out in front of him. Though he hasn't worked that property, it was his turn to staff the company table. He didn't want to sound ignorant if a potential investor came along. "I'll tell them it's great," he finally concedes. Blonde, medium build, with a prominent forehead often set in a scowl, he looks out of place in a suit. In fact, he looks like he *wants* to look out of place in a suit. I've known him since we were thirteen. He was an excellent skier and national champion skateboarder. Competitive, comfortable in the outdoors, good at academics, he seemed like the ideal geologist—only he wasn't left-handed. There is a theory among geologists that left-handers are better at envisioning three-dimensional rock forms. I asked Baknes about this. He shrugged. He said the ability to focus is more important. "The goal is to keep your mind's eye on finding an ore body. In the bush some guys worry about green peppers when they should be doing the geology." He became so engrossed in geology—first at UBC then at McMaster—that he named his pet budgie "Pyrite."

Now he's fretting over other responsibilities: tonight he's getting an award for excellence in prospecting and mineral exploration, named after legendary H.H. "Spud" Huestis. He'll have to make a speech. He'd rather walk over molten lava. The award is recognition of Baknes' work in discovering the Wolverine deposit in the Yukon as well as the wide-open thinking he's encouraged in the industry.

One of the problems with the industry, he says, is that some properties have been "geologized to death." Over the last three decades some properties have been worked over by a half-dozen outfits—each doing soil studies, geological mapping. Why? Because it's easier to raise money on a property with the hint of a prospect than it is to raise money for a never-explored prospect. "But that's the sheep thing," said Baknes. "We try to be a little bit different. Just a little. I'd like to spend maybe 20 percent of our exploration budget on freakish things." Like? He stabbed a finger at a map on the table. It landed in western Alaska. "Don't tell anyone. No one has gone there," he said.

Geologists call the cautious approach that Baknes decries, "brown fields exploration" and the type that he would like to see more of, "green fields exploration." Ironically, it was brown fields exploration that lead Baknes to make one of the largest discoveries in western Canada. The Wolverine deposit in the Yukon sparked the biggest claims rush—as measured by dollars—in the Territory's history. Many geologists go entire careers without making such a discovery.

While people walked past the booth, Baknes told me about the summer it happened. He and prospector Tom Bell were camped in an area that Baknes had worked in before and had recognized from the dead vegetation that something underfoot was affecting the plant life. Such an area is called a kill zone. Lead and zinc in the ground keeps vegetation from growing, just as those chemicals do when painted on the bottom of a boat. "I remember saying to Tom: we could be on a mine right now." A geologist working in the area previously had identified the rocks as quartzites and amphibolites—metamorphic rocks. Baknes recognized that metamorphic rocks are, by definition, made from other rocks. The genesis of the rocks was igneous—a key observation, because the rocks had previously been assumed to be sedimentary, and therefore unlikely hosts for a particular style of mineralization—and the circumstances were exactly what Baknes was looking for.

Further prospecting and a lot of good, detailed ground-work led to a drilling program. There were twelve people in the camp. Only when the drill brings up a core sample can the geologist really tell what's happening below the surface. One day the core came back glittering—a sure sign of mineralization. Seeking some quiet away from the craziness he knew would descend, Baknes pushed off into the lake in an aluminum skiff and drifted around. Within days the remote camp was a hive of activity, hosting promoters and company officials who flew in. They had to feed and sleep forty people. "We practically needed air traffic control," recalled Baknes. Aetna's share price shot from pennies to $6.50. Millionaires were made. The deposit was identified to be huge.

Ultimately, though, the Wolverine ore body proved to have some problems. It was tilted far into the earth, which would increase the expense of mining. The ore included concentrations of selenium, which is both difficult to remove in smelting and expensive to dispose of (though changing commodity prices and new developments in smelting have changed the status of the deposit since that time).

I left Baknes standing behind the core samples and wandered upstairs, where I found Troels F.D. Nielsen, a Greenland geologist with granite-coloured hair and a permanently crooked brow. He immediately engaged in mortal combat with the woman behind the Kitikmeot Inuit Association booth for my attention. Nielsen talked fast and his lips were wet. He gave me an earful about the alluring possibilities of the zinc belt at the top of Greenland. The same belt has produced mines in northern Quebec. When I eased away he thrust a book into my hand. Still talking, he shifted from mining to history. He asked, Did I know that the US Army had a big hospital in Greenland during the Korean War? Or that the harbour at Ivigtut was a staging point for WWII convoys? No, I said. No, no. I think he was trying to convince me that Greenland was an interesting, diverse place. I believed him! But Nielsen was losing out to the allure of musk-ox jerky, which the Inuit girl was handing out. She told me she stands two arm-lengths away from the plate so she doesn't eat it all. I take a piece. It is rich and smoky, like good scotch. I thank her, then circle around like a bluebottle fly and alight on another piece.

Against musk-ox jerky Nielsen and the Greenland belt have not a chance.

In a room entirely of their own the prospectors sell their properties. There is the feeling of entry-level capitalism: something between the back door of a pawn shop and a community flea market set up on a Wal-Mart parking lot. All around is moneyed promotion—Vancouver, along with Toronto, raises about 60 percent of all exploration money worldwide—yet in this room are the properties that seed the industry. The presentations are goofy, without adornment. An eight-and-a-half-by-eleven piece of paper was taped to the wall with a handwritten-scrawled "properties for sale." The hand that lettered the sign is more familiar with a rock hammer than a pen. Promoters Dan Epp, David Forshaw and Larry Riteman were talking with besuited businessmen. The whiff of deals was in the air. Tomorrow another troupe of prospectors would hawk properties.

The prospector is often the first character in the four-stage development of a property from an unexplored area to a valuable and recognized mineral prospect. In the first stage, a prospector comes up with an interesting stone. He probably doesn't have the money necessary to put the stone's presence in geological perspective (ten to a hundred thousand dollars). His story is the stone and how many more like it there might be and how it may well be the tip of something much larger. A junior or major exploration company buys into the story (second stage here) and spends enough money to put the stone in the perspective of a style of mineralization worth exploring for and a regional context where this style of deposit could be found. To raise enough money to carry on, they need to build on the information to convince their managers or investors that their fleshed-out story has potential. And—very important here—the story has to have a better potential than projects being touted by other geologists.

In the third stage of property development, more is learned about a site. Results of research can be good or bad, expected or unexpected. No matter what, though, more is being learned. The story is refined, some previous questions answered, some new ones are raised. Sometimes the story is very compelling (then it is easier to convince managers or investors to kick in more money). No matter what company, or how skilfully the story is put together, after two, three, maybe four years, it is not good enough—the major drops it, or the junior burns out and can't raise more money to keep testing it. The prospect lapses or goes back to the prospector or languishes in the inactive file.

Fourth stage: a few years later, someone else picks up the prospect, looks at the data from a new angle (or in light of a new discovery somewhere else with similar characteristics) and puts together a new story and starts again. Each time this happens, they begin with a bigger base of knowledge but the same drive to solve the puzzle and beat the odds.

In the lecture hall, again, speaker after speaker was delivering sobering messages to the collected executives and geologists: the industry had lost its status. The questions: can it maintain its current meagre status, or will it dwindle further? Can it reach the status it once had? An exploration company executive says that more is spent on research and development in the home security business than on mineral exploration. Another laments the sway of theorists with the Club of Rome, a global think tank.

After a morning of lectures and notes, I wander out of the hotel. A heavy rain has greased the streets and the cabbies are doing brisk business carting suited mining company executives to and from the downtown area—six blocks to the east. I zip up my windbreaker and lean into the wind. Some of these guys will eat and drink heavily for three days and take a total of one hundred paces.

By the time I got to the Jolly Taxpayer Pub, I was damp and hungry. The pub, which managed to combine cozy with a large seating capacity, was midway on the totem of mining entertainment. The moneyed executives went to a thirty-dollar-a-plate café by Canada Place. The Jolly Taxpayer was where the geologists hung out, and the deal-makers whose best days were done. On the screen was British football; Leeds United and Chelsea are tied 1-all. A small but vocal contingent of Leeds supporters cheered their team over a table crowded with sleeves of beer and a heavily smoking ashtray. The table where I sat was round, heavy, and covered with a clear Plexiglas cover under which many, many people had stuffed business cards. More than any clipping file or sentimental speech, there was the chronicle of what had happened to BC's mining industry. On top were business cards for high-tech and tourism, and below that there were business cards for shipping agents. Far below, buried under the midden of other cards, were the cards for the mineral exploration businesses.

So, what happened?

I think there are two forces that have affected the fortunes of the mineral exploration business: consolidation and prices down over time.

Beginning in the 1960s, large mining firms began swallowing one another. The business plan was that the big porphyry deposits required the heft of ever-larger companies to develop them. And that may have been true. As well, however, there were other effects. Larger companies meant that large-scale economics came into effect. Mineral deposits that were once worth developing were left untouched, just as Bill Gates would in all likelihood walk by a five-dollar note. This, in turn, affected the mineral exploration business, because increasingly explorationists had to target areas with large-scale potential.

In addition, ever-larger companies operating ever-larger mines have meant that mineral prices have not only failed to keep apace of inflation, but they have actually fallen. At one point in the mid-1960s, for example, copper reached $3.20 per pound. That led to lots of exploration and development, which, ultimately, led to a decline in the price of copper. Now mining companies face a daunting situation where they are producing cheap copper from fast-depleting reserves, at prices that do not warrant expensive further exploration. And, during the downturn in the 1990s, they sought to decrease their budgets by hacking at the exploration departments—the equivalent of research and development in other industries. When a company wanted to expand production, it did so by buying out another smaller company, or merging with another big company. This practice led to the emergence of three huge players in the global mining business—all of which, it could be argued, are living off the spoils of previous research. They are so big that a mineral reserve has to have millions of ounces if it is to be viable. Many geologists wistfully point out that the practice is unsustainable, but that doesn't do them any good if they are on Employment Insurance benefits at the time.

Some of the standard complaints about this industry dynamic I find at best sentimental, and at worst, snivelling. I come from a business where, in the 1940s, plenty of magazines were paying a dollar a word. Now there are very few, and many pay zip. Too bad. Get real.

What *are* legitimate about some of the concerns, however, are the long-term consequences. Expertise at geology is as much taught by example as it is by book. Typically, a young geology student will work in the field in the summer and acquire the myriad skills it takes to run a camp—helicopter budgeting, prospecting and more—from older geologists. Yet for the last half-decade, perhaps longer, the hardrock exploration industry has languished, and that has meant that students are neither attracted to the studies in university, nor do they have jobs in the summer. The ensuing demographic gap may be the greatest threat to the viability of the industry. Already the collected lore of the prospector is largely lost. To further subtract the knowledge of a generation of geologists is a threat to the industry as a whole.

It is quite possible that in a few years time the boom in mineral exploration that occurred in the 1960s and 1970s will be looked upon as a one-off event, much as the Yukon Gold Rush is regarded. The boom was a function of a new understanding of ore deposits—particularly porphyry copper deposits—and expanding demand for minerals and demographics. People needed minerals. Science suggested new ways to find deposits. A huge amount of capital was available for exploration. Those circumstances in the western world have changed.

Lunch was good but not great. There was not enough salad dressing. I had a beer then a coffee. I called for my tab. The waitress dropped it as she breezed by. It was for $11.63 and she'd signed it. Her name was Amber. I went outside. It was still raining.

When word went around the community where I live that I was working on a book on prospecting, people came up to me and said things like, "You've got to meet my brother-in-law." Everybody's brother-in-law, it seemed, had looked for minerals. My friend Gary's brother-in-law, Phil, had a run-in with the law over a drug deal, and sought a more legitimate fortune on the banks of the Jordan River. He bought a fancy little floating dredge with a hose that sucked gravel from the river bottom. He also bought a wetsuit that was fitted with a nub which connected to the dredge engine. That way, warm engine water circulated through the wetsuit and kept him warm. For two days he worked the river, then something went wrong. The hose kinked but Phil didn't notice. The water in the hose heated to boiling. Then, when Phil worked away from the machine, the kink straightened, and the scalding water squirted into his back. He drove himself to the hospital with skin falling off his back in veils.

One Wednesday I answered the phone and a man said, "I hear you are writing a book about prospecting. I did a bit of that. We should talk." I said, "Do you drink tea or coffee?"

Barry Rouleau is a tidy man with thinning hair and quiet demeanour. If you had to be stuck in an elevator with someone other than your favourite pro hockey player, he would be okay. He said that in 1968 he was working in the Gold River pulp mill, doing a job at age twenty-four that most men waited for until their mid-fifties. He and his wife bought a nineteen-foot travel trailer, hooked it up to his El Camino and headed north.

Whitehorse in 1968 wasn't the cramped, bitchy town that it is now, full of bureaucrats sniffing the air for hints of tobacco smoke and itching to impose bylaws against mowing your lawn on Sundays. It was wide-open, in a terminally good mood. The day after they arrived Barry's wife got a job with the government. Rouleau soon picked up work washing windows, fencing and swamping trucks for the Keno Hill mine, and he made contacts. He took a job cooking on a Cat train near Old Crow, building roads for oil and gas exploration. The thermometer, which went to minus 68, was frozen solid. It was too cold to work, so the men stayed inside. Warmth was provided by oil-burning pot stoves. Even when he left a pan of water over the stove indefinitely, Rouleau couldn't get it to boil—they ate hot, watered beefsteak. Stuck with them in the trailer was the camp dog. The men, bored, sliced meat from their plates and fed the dog so much that it got constipated. It couldn't have gone outside anyway, or its feet would have frozen.

Hanging around the airport the next summer, he was hired on the spot to cook for a crew at Asiak Lake. It was a haywire operation and he lasted two weeks, but it put him in mind that catering was profitable. He moved to Dease Lake and, with his wife, started Mountain Top Catering. He built a log home. His work put him in touch with area prospectors. He worked at different jobs, on the railway grade or on the highway bridge across the Stikine. Rouleau's job was to stoke fires that kept the concrete from cooling too quickly. In the winter he partnered with Miles Bradford and ran a trapline. (Bradford's wife Cherry is daughter of legendary guide George Dalziel.) They had a Super Cub on skids, and they would go out for three days at a time, living in tents. They trapped martin, lynx and wolverine. Their range extended west and north as far as the Yukon border. "Wolves are easier to shoot out of an aircraft than to trap," he said. "In a bad winter the wolves used to leave the valley and go up the mountains after their sheep. The sheep were the guide's bread and butter."

At that time there was in Dease Lake a ramshackle hotel ran by Jack McCallum, who operated the local weather station and was in charge of highway maintenance. Rouleau started a café in the hotel. It served one meal a day, so there wasn't a lot of hassle with menus.

Rouleau's friends in Dease Lake, Gus and Nelly, had a claim near Dawson, and their friend Jimmy Stewart caught the bug while visiting them. Rouleau went up to visit and ended up minding a neighbour's claim. He was allowed to keep what he could pan from it. "Placer mining is a neighbour thing," he said. "It requires trust." Returning the next year, he worked for a fellow named Crawford, who had a bench claim on Bonanza Creek. Rouleau was paid in gold: half an ounce per day. Crawford would give him a pan heavy with gold and tell him to pluck what he was owed from it. Rouleau learned to weld, picked up skills that serve the placer miner well. He kept as much gold as possible. He sold six ounces to a friend in Cassiar who wrapped it in meat and put it in his freezer. The rest he sold when gold peaked at eight hundred dollars per ounce.

Through his contacts in the area, he discovered that a claim was for sale. Osm and Tanya Juso, a couple of aging Finns, had had the claim for years, but were more interested in picking mushrooms than digging for gold. Their claim was at the top of Goldbottom Creek, which was off Hunker Creek, which ran into Klondike Creek. Water flow in the Goldbottom was so low in the summer that there was only one hour's worth of sluicing per day. Barry bought a claim for ten thousand dollars off an old guy with a hernia. The property came with a Michigan wheel loader with a half-yard bucket and a primitive thirty-foot sluice box. The gold in the creek, he said, was round and heavy, not flaky. In all, there were fourteen claims covering one mile. If the old timers had mined well, there would be no gold for those five hundred feet. Then, where their claim had ended, the gold resumed.

A fellow named Brastbury had worked the river well to Rouleau's claim, so everyone thought there would be no more gold. Next spring, Rouleau bought an old Cat. That year, and for years afterward, he took out a couple hundred ounces a year. There were about one hundred days of sluicing—from June 15 to the end of September. In the off-season he returned to Dease Lake. He disliked the work but his wife enjoyed it. Her job was to pick rocks from the sluice box. "Every hour for me on a Cat is like a day," he said. "When you run junk, every time you hear a sound it could cost ten thousand dollars." Early on while he was learning, he'd spend days building a dam only to have it washed away in minutes. Thinking about where to build a dam, his stomach turned to knots—something called "the first cut." The first time he got such a knot he had to open a bottle of whisky and take a slug before he could figure it out.

Working at the end of the road, Rouleau often had several hundred ounces of gold on the property by the end of the season. He divided it and hid it in tobacco cans here and there. One year he came back to find he'd left a can full of gold behind. Rouleau mined the creek until 1985. Two years after he sold, he heard that the new owner found twenty-five ounces of gold that Rouleau had hidden. It was in a shampoo bottle.

Ariadne Hiller was another prospector I met, the only woman directly involved in the industry I came across. When Ariadne first went to work as a geologist in her native Panama, the company geologists gave her difficult and distant ground. They didn't want women succeeding at what had been a man's profession. "So I started at six instead of seven, and if they were back at four I'd come in at five," she said. "Then in the evening, to prove I was a woman, I'd paint my nails."

I met Ariadne in Prince George, in the offices of Falcon Drilling. She was a fast-talking, forceful woman with big, kinked hair that moved when she talked. Her hair looked like an upside-down champagne glass. She is married to Bruce Hiller.

She discovered a love for geology when her mother, a schoolteacher, left a textbook open to a chapter on geology. There were no universities offering courses in geology, so she was bound for civil engineering. Then she spotted a little ad in the newspaper for a geology scholarship in the Communist bloc. She wrote the exam and two weeks later was in Bulgaria. She took five months to learn the language, then five years to get a degree from the then Institute of Mining and Geology. She specialized in exploration geology—especially copper porphyry deposits. She graduated in 1987 and went to work for Geotech International on a gold property on the Azuero Peninsula.

She quickly recognized an entry-level way of identifying areas of prospective mineralization. The names locals used for landmarks often indicated the rock types: *cerro colorado* was the name for a red hill; *peña blanco* was a white rock; *piedra de amolar* was a rock used to sharpen knives. "I looked for names with white colour or hard rock—they can show silicification in the middle of zonation. These are positive topographical features that may indicate gold." At this time the geological information on Panama was fragmented: the oil and gas people had some information, the canal geologists had other. There was not a centralizing place.

Using basic UN surveys, Ariadne and a partner started amalgamating the information. Then the government stopped the paycheques. She went to work in South America and saw lands that were beautiful. Of Bolivia, she says, "Imagine a place that time has stopped. You have an immense blue sky. There are llamas and potatoes. That is all. That is the definition of solitude."

Ariadne met Bruce Hiller in Panama in April of 1992. Hiller was there on business with Falcon Drilling. They dated but Bruce soon worried about the risks of her job. "He said, 'You are going to get killed one day.'" Soon afterwards she took off from Panama City in a helicopter piloted by a man with what she called a "Hawaii Five-O." At low altitude the engine quit. They autorotated into an urban park, setting down heavily between the swings and the sandpits.

There are the men and women in the prospecting industry. And then there are the dogs. Dave Haughton always makes sure his mineral-sniffing dog, Jason, is downwind from hidden rocks. If there is only a slight breeze blowing, he'll reach down, pluck grass and toss it in the air. Haughton, in his sixties, straight-backed, straight up, clad in sturdy jeans clasped with an over-long belt and ironed work shirt, has trained Jason to sniff out mineralization. With Haughton's training and Jason's nose they make for quite an exploration team.

Haughton lives in a plain house in a field in Saanich, just outside Victoria. He trained as a geochemist and studied rocks from Sudbury which are thought to have leaked from the mantle after a meteorite slammed into the planet. When astronauts brought back rocks from the moon he was asked to help analyze them. When he retired from the Saskatchewan Research Council he wanted to stay involved in mineral research. And he remembers a time, years before, when he'd been working at a lake in northern Saskatchewan and a Twin Otter float plane landed and a man and a dog got out. "The dog jumped off the plane, ran up and down the beach and came back clutching a chunk of covellite—a copper sulphide. I was impressed," he says. "Though that was the only sulphide they found that summer."

The man was with Woolex, a Vancouver exploration company that was involved in an experiment run by the innovative Harry Warren, a geologist at the University of British Columbia. Warren had heard of the Finns and Russians using dogs to sniff out minerals. They claimed that dogs were able to sniff sulphides at a rate of one to five. The Finns ran field tests, pitting an ore-sniffing dog against a veteran prospector on a three-kilometre-square field. The dog, Lari, found 1,330 sulphide-bearing boulders, some of which were buried; the prospector found 270 surface boulders. Later, Lari found a copper ore body and was rewarded with four frankfurters.

Warren put an ad in the paper for dogs that liked rocks, and ended up with a couple of cop dogs. He hid rocks around the UBC campus and let the dogs go. They retrieved enough of the rocks to convince him that the idea had merit and he put together a syndicate of major and junior companies to sponsor a bigger program.

"Dogs can smell much better than an instrument and are far more efficient," Haughton said. "A lot of people thought we were crazy."

One hundred and fifty dogs were put to early test, and the best three rock-sniffers selected. One of the best was Jai, which means boy in Japanese. A mixed-ancestry dog, he was found living with eleven hippies in a big house in Kitsilano. Jai sniffed out sulphides and minerals like you wouldn't believe, attracted to the ambrosial scent of sulphur. Furthermore, his partner liked working with him. Said Al McKillop of North Vancouver: "Jai is the best partner I've ever had in the bush. We never have an argument, he can do everything but operate the can opener and on lonely, cold nights in the BC and Yukon bush, he's mighty fine company at the bottom of my sleeping bag."

The program lasted for five years. During that time the dogs were put to use on Texada Island, in northern BC and the Yukon. They never did discover an ore body and that is probably the biggest reason why it languished. But there were problems with the handlers, too, who tended to be prospectors first and dog-handlers second.

Dave Haughton knew about the ore dogs and their problems. But he liked dogs, had worked with Labrador retrievers in Saskatchewan, and thought that a dog could be a helper as well as a companion in exploration. He ran a test on his brother's Labrador retriever Magnus and was pleased enough with the results that he decided to train a dog. He picked the German shepherd because it had a proven record of training and following scent.

To train Jason he put sulphide-bearing rocks out on the lawn. He used a reward system. As Jason progressed, Haughton hid rocks under straw and lawn cuttings. One day Jason nosed into an area where Haughton had not hid anything. Jason pawed around and came up with a gossanous rock. "I really knew he had gotten into the game." That rock is in Haughton's house. Eventually he moved beyond the yard, and staged searches in a local park. One day Jason unearthed several rocks that Haughton had not hidden, and one in particular. "I looked at it—it had a speck of sulphide in it." Jason was ready.

After an experimental session at Bedwell Bay, Haughton had an idea of what Jason could and couldn't do. He was exploring an area around Harrison Lake. Haughton had identified a self-potential anomaly (that's a geophysical anomaly). Jason found the boulders. Haughton found the mineralization. It was in an area of moss and shade. "He dug at the roots of trees where I couldn't get to."

Once dogs start working in an area of large mineralization, they can become overwhelmed by the sulphide smell. Said Haughton, "Dogs lead you to the area, not the target. And the dogs don't work if the handler sits in a lawn chair and lets them loose. They have to work systematically. It's always a partnership situation."

Haughton had an idea that the nickel-bearing rocks that the Giant Mascot mine had tapped ran northwest to Harrison Lake. Haughton explained, "I found a self-potential anomaly, but no outcrop exposure. He just went ape. That confirmed to me that the potential was good. That it was an extension of the outcrop. We were on the edge of the anomaly. The moss was like a carpet. I couldn't see the outcrop anywhere. So I gave him the command to search. It was as if a column of smoke had arisen. He started digging and came up with a football-sized rock with sulphide in it."

I asked Haughton about problems with dogs and bears. Many people think that dogs are dangerous in the bush because bears will go out of their way to track and kill dogs. Haughton disagreed with the idea. At a camp in Duddridge Lake in Saskatchewan he took a Labrador retriever into camp with him. "As long as the dog was in the camp we never had bears coming around. One day I took the dog with me. The night we left a bear came in—right into a tent. We had to helicopter in a shotgun. They shot the bear, right in the tent."

We talked first in his yard, then over by Jason's compound. Jason has developed pannus, an affliction that makes his eyes sensitive to sunlight. Jason lives in a chain-link run covered with blue tarps. He's a good dog. When Haughton let him out, he identified some old dog treats in my pocket and the scar on my left knee. Though he is four he still is quite puppyish. He bites his tail when he walks and rolls on the ground.

Haughton and I had planned a test for Jason. I would hide rocks of different sorts around the yard, and then he would set Jason loose. He cautioned me that Jason had not worked in fourteen months.

We locked Jason away. From Haughton's garage we fetched two rusty rocks. I buried one under a pile of granitey-looking rocks near the house. Then I hid the other in the garden. Haughton insisted that he remove himself from the process and shut himself in the garage.

I walked around the house and tucked the rock under a rhodo, and covered it with needles and dry earth. Then I walked back around the house and Haughton let Jason loose. That's when Haughton bent down and flicked some grass in the air. Jason nosed out the first rock so quickly that it hardly seemed a test. Mouthing the rock, proud, he strutted around the yard, ignoring Haughton's good-natured attempts to get him to drop it.

For the second part of the test Haughton asked me to stand at a distance. He marshalled Jason over to the west side of the yard and ordered him to search. At first the dog was distracted by a fledgling that had flopped from a nest. When that was cleared out of the way the dog swept the ground, looping like loose handwriting. Back and forth he went. Jason was so resolute on the hedge that Haughton finally said, "Is it in there?" I said no.

He directed Jason to the front of the house. The dog made one, two trotting loops, then honed in "as if there were smoke from a chimney." It was as if his nose were on rails. Haughton pried the rock from the dog's mouth but Jason did not want to stop. He pawed around in the garden and found a rock with the slightest sericite facing.

Haughton tried to get the rock from his mouth but Jason would not have any of it. He bounded away, came back, and bounded away again.

III Prospecting School

The mail-out for the British Columbia Institute of Technology's prospecting course said to be at the field school in the hills above Oliver, in the Okanagan Valley, no later than 6 p.m. It was May. Most students were there two hours early. We parked our vehicles—older-model Fords and Chev trucks, mostly, dirty and brimming with camping gear—on a tufty lot bordered by buckskin logs.

The field school is in a little valley, backed by a sparse forest of Jack pines and fronted by unprosperous-looking farms. The camp is four large plywood buildings painted a shade of green that you just know was donated. Set randomly between the buildings and pines are 1970s Arctic oil exploration-era metal trailers. Beefy and futuristic, they were designed to be towed by bulldozers across frozen tundra or ice.

Two women taking the course were setting up in one bunkhouse, the rest of us merged outside the door of another. I asked a burly middle-aged man with sunburned cheeks and cheery voice where I'd find a bunk and he stabbed a thumb over his shoulder at an open door. Inside, I found an empty room, eight by ten, bunk beds on either side. The only luxury was a hook on one wall, so I claimed it. If four people were to sleep there you couldn't get any tighter unless you were in a submarine.

"Any room in the hotel?" Dan Wilson, clutching two large duffel bags, had just driven in from Vancouver. He was in his ageless fifties, bearded with prominent teeth and lively half-moon eyes. He carried himself in a permanent slouch, not from the shoulders, but almost athletically, from the hips. That posture was explained, perhaps, by his line of work; he drives long-haul transport. He hauls cargo of all sorts: lumber, pipe from Salt Lake to Alaska. He drives ten thousand kilometres per month. His current truck was seven years old and had 1.35 million kilometres on it. He works from 7 a.m. to midnight, keeping alert by listening to CBC, NPR, scratching down names of jazz records on a little pad he keeps at his side. When the road turns Kafka on him—when the asphalt claws into the cab—he'll pull over and do something else for a few hours. If he's near a town he'll explore graveyards—the graveyard in Whitehorse, he said, was excellent—but more often these days he took the Estwing rock hammer he kept in his cab and pounded at a nearby outcrop. He said he didn't know what he was looking at but he wanted to find out. That's why he was taking this course. We held the first committee meeting of the room, and decided by a 2–0 vote to close the border to further immigration. In an unwelcoming gesture we pitched our extra gear on the overhead bunks.

Outside, someone had lashed an orange tarp between trees. It was tied off by yellow rope and held aloft in the middle on a pole with a sock rolled onto the tip. Under the tarp, tables were crowded with fist-sized rocks. They were infused with an orange tinge from the tarp. Several men stood around the displays, clasping the rocks. Patrick Monk was ahead of me. Professorial and grey-haired, he hoisted a milky rock, tilted it this way and that. "Do you know what kind of rock that is?" said a large, bearded, reptilian-looking man. He was perched behind the table on an upturned plastic bucket. I looked at his name tag: Tom Richards.

Since talking with Tom Bell and Pat Suratt, I had come to realize that "Dr. Tom Richards" (as Richards is sometimes known) is a storied figure in the strange sub-subculture of BC exploration geology. An enormous man with a leathered face, ropey hair and a two-pack-a-day voice, Richards runs an exploration company in the summer and in winter hosts what are billed as seminars in prospecting but are actually sermons on evangelical geology. He thinks that geology, not hockey, is Canada's true pastime. Again and again I heard the legend about how, in the late 1970s, and with only his trademark plastic buckets of samples and photocopied handouts on petrology, he tutored a bunch of back-to-the-landers, skanky bushmen and cowboys in the ways of rocks. His group, Swami Tommy's Revolving Ore Bodies (STROB), became to mineral exploration what the Trail Smoke Eaters were to amateur hockey; their influence and ability extended far beyond the valley. Richards' protégés are regarded as the best prospectors in the province.

I had heard so much about Richards that I had concocted an image—of a sinewy mountain-trotter, effusive about all things geological—that was completely at odds with what I saw. Cigarette hanging from his lip, big-bellied, he resembled Gandalf in *Lord of the Rings*.

Patrick Monk studied the rock and said, "Calcite."

Richards said, "Good guess. Now prove it."

When Monk hesitated, Richards heaved his considerable

frame from the upturned five-gallon pail he was sitting on and hoisted another remarkably similar-looking rock. "Weight. Heft," he said. Other students put their rocks down and gathered around him. "Feel this rock and that rock. One is heavier than another. That's called specific gravity but the name isn't as important as the idea. Some rocks are heavier than others." He said that anything with a specific gravity of more than 3, such as fluorite, is noticeably heavy for rock and something like galena, with a specific gravity of 7.6, would be strikingly heavy. I elbowed past Monk and took the calcite in hand. It felt lighter than an average rock. Calcite has a density of 2.6, or 2.6 times the weight of an equal volume of water. A similarly sized chunk of barite is heavier: 4.5. "The numbers are not as important as the relativity," said Richards. "Get to know the rocks—how they feel." Self-consciously, we stood around the table, hefting rocks and looking thoughtful. Everyone had a faraway look on his or her face.

Dinner was chicken on pasta. We ate cookhouse style, stabbing buns at our plate, wielding the cutlery like heavy equipment. Around the slops pail men said, "Excuse me." They drank coffee like it should be served by the pint. Afterwards, the head of the school spoke. Rob Stevens is a curly-haired geologist with a big-block IQ. If he didn't have a scruff of beard on his boyish face he'd risk getting ID'd at bars. He introduced the school staff: himself, Tom and geologist James Moors. Then we introduced ourselves: two women, Kim and Dot; a raft of one-syllable names—two Toms, two Dons, two Dans, plus a few Patricks. There wasn't a twenty-year-old in the crowd; the average age was mid-fifties. They included truckers, a lawyer, an accountant, a carpenter, a tow truck driver, a locomotive engineer. Gary Kmett was the only American. "You all talk funny," he said. Tall and with a high forehead, Kmett teaches geology to fourteen- and fifteen-year-old Navajo kids at a high school in Tuba City, Arizona. Bored out of his tree one afternoon last term, he went online in search of a summer activity and found that the Prospecting and Exploration Field School is the only one of its kind in North America. So he signed up.

Even in the momentary introductions the characters settle out: Kerry, a logging truck driver from Lillooet (and the one who showed me into the bunkhouse) said he used to pan agates in the Fraser River with his dad. Don, a jailer on a reserve in northern Alberta, introduced himself with a lengthy talk on a placer claim he has around Barkerville. I think: "Jesus Christ: he's come to talk." Dot, from Port Alberni, who was at the prospecting school last year with an eighty-one-year-old fellow rockhound is the only student attending for purely recreational reasons. "I'm a repeat offender," she said.

These people paid $750 to attend the week-long course. Even so, the course is subsidized in part by the British Columbia and Yukon Chamber of Mines, which represents the mineral exploration industry. The BCYCM is trying to counter a decade-long decline in the number of active prospectors. As low-cost talent scouts in the industry, prospectors scour the province for areas with mineral potential, stake the ground, then sell or option the claims to exploration outfits who, in turn, apply more rigorous and expensive exploration techniques—stream sediment geochemistry, soil geochemistry, geophysical surveys and, finally, drilling—to the properties in hope of interesting a cash-heavy mining company. The first part of this process—the identification of areas with mineralization, the mapping, interpreting and staking—are the substance of the two-year-old Prospecting and Exploration Field School.

After dinner Richards lurched to the front of the room, stretching his face as if the skin was too tight. He clutched a bamboo pointer with a shattered end. If this was hockey school, it is where the instructor would give us a tone-setting anecdote about playing hockey alongside Kevin Lowe, the importance of weighing-in for teammates when knuckles fly. The idea would be to signal the school's attitude. "Rocks are something I've grown quite fond of," Richards growled. "I can talk to rocks and listen to them." He was educated at UBC, and studied under H.H. Reed, a venerable geologist whose motto was "He who has seen the most rocks is the best geologist." As a grad student Richards had an office beside Harry Warren, the classic old-style geologist who used mineral-sniffing dogs at UBC. "Harry Warren was able to look at a rock and be interested in it. Modern geologists aren't," he said.

After graduating from UBC in the early 1970s, Richards worked for the Geological Survey of Canada, where people who had "a passion for rock" surrounded him. After a decade he left the GSC to get into the basics of the industry: prospecting. But the transition wasn't easy. Richards says he had to break training when he left the GSC to become a prospector. "A geologist's knowledge is only good for identifying the right area to look in." That's not the same as identifying a promising rock.

According to Richards, there are three types of prospectors. The first are "Those who break everything in sight. You can always find them." The second type are "thinkers, who reason their way around." The third are the prodigies. Said Richards: "They can walk off the trail and find copper. They have a natural nose for it."

During a break in Richards' lecture I walked outside. The air was chill and brittle. Light from the stars seemed to drill down toward us. Richards' comments made me think of those people always on the fringes of orthodoxy who claim to have a special ability to detect minerals. In the 1960s a prospector working in northern Ontario said he could read the Earth's secrets from a colour television, which he plugged into the ground. And in Tasmania years ago a prospector said he could dowse minerals by placing a sample of the desired mineral in a little container on his dowsing rod. Skeptical researchers with the Tasmanian department of mines buried twelve different minerals on a property then invited the prospector to find them. He located eleven of the twelve sites.

None of these nor the many other feats of mineral finding compare, however, with the alleged talents of a willowy, fine-spoken woman who appeared in the Okanagan Valley in the 1930s and made a great promise to the area's drought-stricken farmers. If they paid her a fee, Evelyn Penrose said, she could find them water. She had located well sites in Hawaii and was prepared to do the same in parched south-central BC.

Farmers were doubtful for dowsing had not worked, nor had they heard of women dowsers. Yet Penrose never questioned her talents. Walking with an orchardist through his drought-stricken operation, she was accosted by an unseen force. As she recounted in her memoir: "I grabbed his arm to steady myself. 'Water!' I gasped. 'Lots and lots of water!' He looked at me in amazement, obviously thinking it impossible that there could be any water in a spot that he knew so well, and over which he walked every day of his life. I followed this powerful underground stream with my divining rod to a little wood by the side of the lane. Here I found the intersection of two underground streams which made the reactions stronger than ever." A well was dug on the spot. It produced thousands of gallons daily. Henceforth she was known as The Divine Lady.

Most of what is known about Evelyn Penrose comes from her autobiography, which of course makes it suspect. Yet in her time she achieved a hero's status. She had the ability, it was said, to read the Earth's subsurface like a trapper reads an animal track. Soon after appearing in the Okanagan she was appointed the provincial government's first official water diviner. And soon after that she began doing something even more remarkable—divining minerals.

Penrose was born in Cornwall, England, to a line of water diviners. Her father inherited "the gift," as it was known, from his mother, who trotted around her Cornish village clutching a wooden rod. Her father, Penrose recalled, could dowse at will with fresh-cut forked hazel or willow. "I have seen the rod skin the bark off itself and sometimes twist itself into a sort of rope in his hands, but nothing in the world stopped it from turning," she recounted.

Penrose discovered a talent for detecting underground treasures while visiting California. Travelling through an open field, she experienced a much stronger feeling than she had when dowsing for water. She jumped and twisted, then felt nauseous and developed a stunning headache. She was told she was near a natural gas deposit.

Penrose didn't fit people's idea of what a prospector should look and behave like. Skinny as a marathoner and with a modest overbite, her appearance was of a feeble woman who might retire to bed with the slightest headache. To prove once and for all that her talent was real she staged an elaborate experiment in a Victoria hotel. A wood table was placed on a rubber mat, and each leg placed on a plate. A chair was similarly set up, with each leg on plate and rubber. Penrose—clad in rubber bathing cap and rubber boots and cinched into a rubber mat to protect her from electrical fields—sat at the table. She was handed a map she had never seen before. Swiftly, she pencilled circles around three areas. The map was handed to mining experts. They were stunned at what they saw: Penrose had identified areas where minerals had been found.

Later still, Penrose was contracted to find water for reserves in the Peace River country and there she found oil. "I stand quite still, stretch out my arm and turn my hand so that the palm and fingertips point upwards and act as a radio receiver. I keep my hand gently moving sideways and backwards and forwards, and turn slowly round. When my hand gets into line with the oil, water or mineral, I immediately feel as if [there is] a little thread coming out of each finger, connecting me with the deposit. This little thread becomes a string and then a rope and, unless I break the contact by running my left hand down over my arms and fingers, my arm will nearly be pulled out of its socket."

So strong was this pull that, Penrose claimed, she had once come across side-by-side mineral deposits and very nearly been yanked in two. With each hand attracted to a separate field, she was torn. She saved herself by dropping to the ground and burrowing her hands into the earth.

Penrose left BC and travelled to France, England, Rhodesia, Jamaica and Chile. In Australia she discovered another talent: identifying people's illnesses by circling them with a pendant. When her pendant swung, she had found the troubled area. The police were so impressed they asked her to deploy her skills at finding criminals but she declined. She said it might get her killed.

At 10 p.m. we stumbled out of the cookhouse/classroom and into the bunkhouse. All of us were dumbfounded to realize that this was a credit course, with credit course expectations: homework, assignments, and classroom attentiveness. The bunkhouse was unheated and the night was cold. I climbed into my sleeping bag quickly. I went to sleep thinking about two things: one, the weight of calcite; and two, that if there is a next time, not to take a room near the crapper.

Day two. Dave Watson had no trouble sleeping. "I dreamed of rocks!" he said in the morning. Model-handsome and in his sixties, he is snow-haired and trim. He drove in yesterday from Banff in a camper van—one of the few student vehicles that look of the middle class. He's as tidy as a roadside park. Before he retired Watson was a highway engineer. He helped design Banff's riparian underpasses and overpasses. The overpasses, he says, were cosmetic and a waste of money. He comes to an interest in rocks via what sounds like sibling rivalry. His brother is a geologist working on a gold property in the Maritimes. Watson says he signed up for the prospecting course so he can make good investment decisions but I doubt it. I bet his brother used to give him Dutch rubs and now Watson wants to get him back. He speaks with a hesitancy that is difficult not to be impatient with. Maybe he spent too long holding his tongue in bureaucratic meetings. He's brought along literature on opals. Smiling beatifically, he set his plate of steaming eggs and hash on the table, leaned close, fixed me with his blue eyes and said, "I think there should be opals in BC. If you look at the geology around Revelstoke . . ."

He's been to Australia several times to study gemstones. There are so many opals to be had in Australia that people hunt them on weekends. It is free for all now, but until 1992 casual opal hunters needed a wonderfully titled "fossicking licence." Who wouldn't want that in their wallet? In the Australian town of Sapphire, Watson visited a store that sold sapphires. Beside the store was a hole three feet wide and forty feet deep. In the mornings, the store owner descended a ladder into his own mine. He jackhammered until lunch, scrubbed up, and worked in his store as a jeweller in the afternoon. Watson was smitten with this idea and I could see why. The Australian was a one-man De Beers, incorporating the work of thousands into his workday. Watson said he knew of many places around Banff where opals and sapphires might be found.

After breakfast Rob Stevens distributed big blue binders embossed with the BCIT logo. They contained our schedule for the coming week. Mornings were to be in class—i.e. in the mess hall—going over academics; afternoons were to be in the field; and evenings . . . well, it wasn't clear what was going to happen in the evenings, but it didn't look like we'd be playing blackjack.

Stevens talked about maps. He paced as he talked and paused often enough to let us know that we were leapfrogging over subjects worthy of entire third-year university courses. He said geologists are, at heart, map-making creatures. Their instinct, when first hearing of an area, is to pull out a map, or to call one up on their computer screen. When they walk a prospective property a map is always at hand. Their work largely consists of creating maps. They begin with topographical maps, which include landscape features and elevations given as contour lines, but amend them to show the surface rocks and what lies beneath. To do this they use all sorts of techniques to indicate the angle at which various beds outcrop. The most important concepts are strike and dip. The dip is the angle at which an outcrop is inclined from the horizontal. The direction of the dip on an exposed surface can be found by watching which way water runs down it. If water runs north off an outcrop, then it is said to dip north. Strike is the line that a rock bed makes with a horizontal plane (often thought of as the waterline that would be formed if the bed dipped into a lake). "Strike and dip, strike and dip," he intoned. "Get that in your heads."

It occurred to me as Stevens talked that there was a simultaneous macro lesson going on. It was in the pedagogy of wooden benches and nicotine. By 9:30 a.m. the students were shifting; some had opted to stand. Several others had slipped out to smoke near a doorway. I sensed that this was an action-oriented group and that our collective temperament was at odds with that of the cerebral Stevens.

At noon we hustled into a big white van. The variety of equipment stuffed in the back was amazing: new packs, fluorescent vests, new rock hammers, boots, runners, water bottles, walking sticks. We bounced down off the slopes of the mountain and turned south on Highway 97. We passed sunny little rectangles of orchard, lithe farm girls in tank tops, everywhere "Danger: Fruit Spraying" signs. So many of BC's resource industries were beleaguered—mining and forestry came to mind—that to see such prosperity was unusual. There's a 1950s *Beautiful British Columbia* magazine aspect to these farms, as if the father is always working in short sleeves, the farm workers are cheerful ethnic characters and at the end of the driveway, a pigtailed daughter is peeping into a metal mailbox in search of the latest 4-H newsletter.

Stevens gunned the van up a wide suburban road leading

to Mount Kruger, a blocky unimpressive hill on the west side of Okanagan Lake. The wit-chit-chit of front yard sprinklers was audible even over the van's labouring engine. Suburbia was rising up the slopes like a floodtide. Just beyond the last house we pulled in under a bluff. While we wiggled into our packs, Stevens gave us the Coles Notes on the area: it is part of a batholith, a deeply formed mass of igneous rock that has intruded the existing volcanic rock. The intrusion was accompanied by all sorts of geologic heavings, during which time mineral-rich fluids intruded fissures and cracks. The particulars of the mineralization—gold, silver and copper hosted in limestone—made it a skarn deposit.

Exploration on the property began in 1901 and it was worked as a producing mine intermittently for three decades until it was closed for good in 1940. Though the property has been explored and drilled since then, its primary purpose seems to be as a rehearsal ground for UBC geology students who, like us, come here to practise note-taking and work on basic rock identification.

I'm paired with Don Chimco, in his mid-fifties, who said, "I'm coming at prospecting from reverse." He recently retired from a profitable scrap metal business in Kitimat. Now he wants to find metals the way nature created them. He is white-haired and athletic, one of those self-confident men who are completely without pretense. It was hot. The students fanned across the hillside, nervous, self-conscious, tap-tapping at what Stevens calls float. Float is the general term for rock fragments that come from an outcrop. "Is it really okay to bash rocks?" we all seemed to wonder.

Students peered at specimens through their hand lenses, then scratched observations in their notebooks. The land was green but in weeks it would bake brown. Chimco and I seemed less eager to mount the hill than the others. We took our time bashing at a big float near the road. I was overwhelmed by the sense of openness. After the closed-in Coast, where to enter a forest is to have a fistfight with salal and brush, the green open hills looked appealing.

It took forty-five minutes to make our way to the top of the bluff. We stood with our thumbs hooked into the shoulder straps of our packs, peering at a hand-dug open pit. That, said Stevens, was the former mine. What we saw was so at variance with the prevailing notion of a mine that it did not seem to warrant the term. It was more of a protégé mine, a pocket excavation. Yet from this mine a total of 111,252 tons of ore was won. It looked like something a Christian pillar saint would live in, or where US Delta Force troops will gun down Osama Bin Laden. A roof of rock was supported by three rock pillars. All the other rock was mined and milled. A shopping cart had made its final descent and lay upside down amid food wrappers, bottles and broken chairs. Just metres from the expanding edge of the town, the old mine site is a sort of activity centre. Two turbaned East Asian teenagers circled the summit on an olive ATV, waved at us and vanished down the slope in a trail of dust.

Chimco and I stepped away from the workings and stood over a dozen fellow students sprawled around a doughnut of broken rock. Chimco and I had made a slow ascent and simply felt the heat. We were warm but not tired. It was a shock, then, to see several students pale and panting. "I think he is going to die!" I scrawled in my notebook.

Stevens split the class in two. I was in a group fetched to a rock face on the far side of a draw.

We slid down the draw whooping, then worked our way

up a two-steps-up, one-step-down talus slope. Talus is broken rock, usually angular, lying at the base of a cliff or steep slope. The talus sounded like broken glass. The bluff faced east and in the awning of shade, resting with his back to the rocks was Richards. He looked like an aged Tom Sawyer. How he got there I could not understand. He seemed to have trouble moving around the mess hall, yet he had powered up this moderately difficult climb. I think it had something to do with the pine stick he carried. He got antsy when it was not at hand.

"Look at the cliff behind me," he growled. It was layered with different shades of black rock. The trick, he said, is to tell which layer contained copper. A chemical laboratory can do it, but so much the better if a prospector can identify copper quickly. He took a vial from his pocket and several old nails. He dropped acid on the rock and pressed the nail onto the moisture. He looked at us like a magician waiting for the rabbit to turn into a snake. He rubbed his index finger in the solution and held it up. It was green. "Copper," he said.

"Rocks are something I've grown quite fond of; I can talk to rocks and listen to them," said Richards one evening. Bamboo pointer in one hand, cigarette in the other, he was about to give what he called "the mosey lecture." The rest of us, aching from a day scrambling around yet another old mine site, were folded onto the hard benches of the cookhouse/lecture hall, pens and paper at the ready.

He said that mineral exploration is accomplished by two methods—direct and indirect. Direct is pounding a rock with a hammer. That's what prospectors do best. Indirect exploration methods include electromagnetic gravity tests, magnetometers and geochemical analysis. Indirect methods attempt to identify phenomena that can lead to an ore body, much as the footprint leads the detective to the culprit. In Canada, the geology favours indirect methods in the east and direct methods in the west. Much of the Canadian Shield, which is layered vertically, is covered by glacial till—gravels. The Shield yields best to the probing eyes of electromagnetism. British Columbia, on the other hand, is in general pancaked horizontally—the topography is 3-D. The challenge, or what Richards called "the art" of prospecting in BC, is to think in three dimensions.

The art of prospecting, he said, is the art of looking. "What the prospector needs to look for can be taught, they can be seen and touched," he said. "They are physical. But *how* a prospector looks is personal." He stabbed the pointer at a picture of orange mineralized soil known as a gossan. A gossan is a décolletage to a prospector.

"Looking is a mental exercise," he intoned. "Looking is a discipline." At this point twenty-four pens posed above twenty-four notepads. This is where the mosey comes in, he said. Twenty-four pens remained in the air. He smiled, Buddha-like.

Mosey? This is the geological equivalent of being cool. If you are, like me, self-absorbed, worried about something all of the time and everything some of the time, the effort required to relax and unlock yourself to geology is remarkably difficult. Most of us defaulted to what Richards called the rock-breaking strategy, meaning you pulverize every boulder. It wasn't true mosey-school style, but at least if you got lost you could follow the rubble back to the school's van.

Day three. A marvellous, goofy, intuitive day. But how could it be anything but after the way it began? I rose at 5 a.m. Padding across the gravel I met Kerry, the truck driver, clad only in shorts. He'd showered and was enjoying the morning sun on his expansive stomach. I, skinny and always cold, was envious of these big, well-heated fellows. In a loose way Kerry reminded me of the wisdom of the Sikh saying that too much education is like drowning. He commented on what is, not what might have been or could have been.

There's a country the size of Belgium back in behind the Cariboo, and Kerry has seen a lot of it. He's hauled logs all over BC. Once, he had a contract to take peewee sawlogs from Anahim Lake in the Chilcotin to Okanagan Falls. A round-trip took thirty-two hours. He grew up in Lillooet and knew famous newspaper editor Ma Murray well enough to say good day to her. He went to see movies in a theatre converted from the gold-rush-era building that housed camels. When he was a boy the town's jail was a building constructed entirely of steel bars; there were no solid walls. Kerry's father had a placer claim. It was something a man did in the 1950s, like pack a rifle in your truck. Only less handy.

Kerry can't recall a time when his father actually worked the claim. He'd point it out as they hurtled past on the highway. "That's where the claim is, son," he'd say.

One afternoon I was paired with a fellow named Andrew—"call me Drew." Tall, with a careworn face, Drew established himself as the camp dandy on the school's first night when he explained his presence by saying his buddy Patrick Monk "dragged me here." Drew and Monk were knocking back drinks at a Commercial Drive restaurant when Monk mentioned that he was going to prospecting school. A party-worn musician and sometime prop-acquirer for the film industry, Drew came along for a lark. “The last time I picked up a rock was to throw it at my younger brother,” he drawled.

Bent at the neck like a sunflower, Drew dozed during morning lectures, or simply walked out. He was outfitted head to toe in clean, pressed outdoor gear: Carhartt vest, stone-washed shirt, green trousers. After a hot afternoon at one mine site he made an unsuccessful representation to Stevens for an hour off. “To have cocktails,” he pleaded.

We drove north on Highway 97 to the Vault claim. Of the four different properties that prospecting school students visit, the Vault is the only one that has not hosted a mine. It is a gold-bearing deposit, and often cited as an example of how a modern-day prospector might profit from an easy-access find. The Vault property is one of those properties that the province’s geology offers up every now and then to entice backyard prospectors into the business. It is located in virtually handicap-access terrain, a ten-minute walk from the highway. Such a property might be optioned for ten thousand dollars plus benefits.

To access the property we had to drive through a farmyard where blossoms rained from the apple trees and the fences were lined with firewood. The house was small and cozy and made you want to move there. The owner, said Stevens with a coy grin, had a habit of being naked when he’d driven through before. We put on our gear in a grassy area at the foot of a hill. A lane led up the hill and was swallowed in forest. We hiked to the brow of the hill and ate lunch perched on a rock outcrop amid pines.

The first part of the afternoon’s exercise was to practise pacing and mapping. But, conveniently invoking Richards’ stated belief that you can’t do geology and prospecting at the same time, Drew abandoned the pacing to a series of good guesses.

So twenty of us walked along, trying to mosey—to, as Richards had said, imbibe the environment, to let the rocks speak to us—and count at the same time.

Whenever I got over fifty paces, Drew interrupted and I forgot my count. After just a few minutes together, I realized he was not just a bad influence. He was a horrible influence. Whatever I had thought of that was bad, he had thought worse. In Oliver, before coming up to the school, I'd spotted a For Sale sign on a marvellously humble hotel, and I had a brief day-dream about buying it and fulfilling a lifelong fantasy of running a seedy bar. Drew had had the same daydream, only he had gone and checked out the price. "Four hundred and seventy-five thousand. But I think you could get it for three-twenty-five," he said. His commentary for the afternoon was of leggy Danes and fancy drinks. He had an especially strong craving for a Manhattan, which in deference to the prospecting school he thought might be renamed Azimuth.

Drew was cagey about his past. He said he was trained at the Royal Conservatory and he played piano for a variety of bands, the longest with the group Sweet Dick, which he'd been with for twenty-four years. For a quarter-century he's made his living in television production. He found a lot of material that was used in the *Andromeda* series. He once had to locate the cockpit of a water bomber for a television series pilot called *Night Flight*. He talked about so many things, and so quickly, that I had trouble keeping track. He said he was reading a book about how ancient Egyptian rulers mistreated the mud people. Then he told me about the time an untalented Irish folk singer was disturbing the atmosphere at his local watering hole. "I approached him and asked if he knew what 'defenestration' meant. The guy said 'No.' I said, 'It means, "thrown out the window".'"

While I was mapping, a bug fell onto the paper and rolled into a ball. I touched it with the pencil tip. Drew watched me and said, "Do you know that Madagascar has the most bugs?"

All around us students industriously beavered at their assignments, studying their compasses. Drew wandered in the woods, head high like a dog sniffing a breeze. Yellow arnica was in bloom; he picked one and slipped it into his lapel. With our assignments—a mass of cheaty notes and generalizations—we arrived in a grassy area marked with a stick protruding at an odd angle. It marked a drill hole. Rob Stevens gave us a brief rundown on the property. Several companies, including Riocanex, Dome Mines and Seven Mile High Resources had drilled over twenty-four thousand metres in seventy-two holes. What we were looking at was one of them. Gold and silver had been discovered, but not in quantities that warranted mining.

We were given a new assignment: to trace quartz float to its source. Stevens pointed to scattered white boulders nearby. "It had to come from somewhere. Find it," he said. Drew and I wandered up the hill, somewhat floppy, as if we both needed naps. Through the woods I could hear the shouted comments of the other students. Near the summit of the hill Drew found a quartz vein, far too narrow to have produced the float. The vein ran roughly north-south. We followed it north until it petered away. Without bragging, I'd say I know about a hundred times as much about geology as Drew. I suggested that since we'd traced out the vein in one direction we should trace it out in another. Yet Drew wasn't interested.

In what I can only explain as a sort of sixth sense, he insisted we explore an area perpendicular to the vein. To me his notion was ridiculous. We walked west up a grassy swath between sparse timber. It was very nearly *Sound of Music* country: open, green and joyous.

"What's that?" said Drew. Ahead of us a massive quartz vein broke through the earth like a whale rolling at the surface. We bent and placed our hands on it. It was waxy and cool, like an enormous toenail.

James Moors, who was nearby, commended us. We were the first to find the source of the float. "Now, where does it go?" he asked. He wanted us to look further, to map and trace and identify.

But Drew would have none of it. That was too much like marriage to him and he was not the marrying kind. He sat with his back against a log and mused, "I was groping this black lesbian singer not long ago . . ."

As I sketched our find onto the map, I suggested we incarnate our talents as the Henry/Drew Group and get listed on the TSX. But Drew wasn't interested. He gazed longingly at the highway, a dark river to conventional pleasures. "We could mosey down there and hitchhike to Penticton," he said. "Slack Alice's is a great peeler bar."

The relationship between the gritty world of prospectors and the hallowed halls of academia may not at first seem clear. But Harry Warren—a geologist, UBC professor, Olympic athlete, biogeochemistry pioneer, and patron saint of the mineral and exploration industry—was far from a linear thinker. Warren, a Canadian Mining Hall of Fame inductee and Order of Canada recipient, made invaluable contributions to the prospecting industry; his discoveries were adopted by mining companies worldwide, and he helped established the very prospecting school I attended.

In the 1930s in the hills near Cache Creek, Warren, then

a young geologist, made a discovery that would ultimately have wide-ranging consequences, affecting everything from mineral exploration techniques to human health to the school of prospecting. Warren's discovery had nothing to do with ores or outcrops and everything to do with himself. The handsome, gifted, athletic son of a well-connected British family whose ancestors hobnobbed with Prime Ministers Burke, Gladstone and Disraeli, Warren belatedly discovered that he hated digging. Perhaps it was genetic. The all-fours scraping and rooting that is the foundation of geological fieldwork was, to use a phrase that Warren might have decanted himself, off-putting.

For a geologist, especially a young one who does not yet have the seniority to claim a place in the lab, an aversion to digging is akin to a basketball player realizing he does not like jumping or a barrister being repulsed by debate. Yet within a single afternoon, Warren recalled years later, he decided the drudgery of digging was not for him. The hiking and climbing and painstaking analysis he could abide. But digging? No.

In his discovery that there was some aspect of geology that he did not like, Harry Warren was not alone. Every year since geology became a profession, young geologists have left the discipline because of some aspect they dislike. For many it is the long absences from home and family. For others it is fear of heights or falling rocks, or the discomforts of bugs. Still others are near phobic about bears, or cannot stand the yo-yo economics of the resource industries. Look through journals like the long-defunct *Western Miner* and you'll find regular articles with titles like "How to Attract and Keep Young Geologists."

But the difference between the multitudes who quit and Harry Warren was that for Warren nothing—absolutely nothing—was such a problem that it couldn't be tackled or worked around. Perhaps that attitude was a result of his Olympic level of athleticism. On the night he discovered that he disliked digging, lying on his back, pinned into his blanket, he wondered about his dilemma. The purpose of digging was to get at the rocks under the overburden. It seemed to Warren that there had to be a solution. This was, after all, the man who as a sprinter patiently waited at the sidelines during one of the most celebrated Olympics in history—the 1928 Games in Amsterdam—then in a meet several weeks later beat the stars of the day and blasted to a world record. Problem: solution.

As he lay wrapped in his blanket, Warren thought of tree roots fingering deep into the soil in search of water and nutrients and trace minerals. Why not use the trees or other plants to do the work of the geologist? It was no secret that soil composition affected plant growth: prospectors had known for years that inexplicable barren patches in otherwise verdant areas often indicated poisonous concentrations of mineral. What Warren vowed to do was apply the evolving parts-per-million and parts-per-billion capabilities of science to the analysis of plants. Thus was born the fascinating partnership between what may be Canada's most eccentric geologist and an eccentric subcategory of geology known as biogeochemistry.

Long before the term "lateral thinking" entered the language, Harry Warren was thinking and living laterally and, as his opponents on the rugby and field hockey pitches could attest, he could turn lateral at full stride. He did many things, often simultaneously, and much of it he did very well.

He was born in Anacortes, Washington, but was thoroughly Anglophile. His father's family was of that mainstay of the fighting class of British society whose members could lay dormant in the estates of England for generations—then, when occasion demanded, shed their tweedy husks and perform great military deeds. His ancestors included the commander of the man-of-war *Triumph,* which played a crucial role in the defeat of the Dutch fleet at Solebay in 1672; Sir Edward Verney, who forever earned the sympathies of the British Royal family for slapping a mouthy priest in the face in the presence of Charles I, then fighting the parliamentarians at Edgehill; and an uncle who was one of the first to die fighting Germany in 1914. A tributary genealogical stream included the Dowdings, who took their name from the Saxon words *dow*, meaning to be able, and *ding,* to strike.

Even when they weren't fighting for King and Empire, the Warren clan seemed to be at the fore of whatever they did. One ancestor loaned Edmund Burke twenty thousand pounds, launching his career in politics, another partnered with Benjamin Franklin in a brewery and yet another was a noted ventriloquist, able to make trees and furniture babble in conversation. Even before he made a name for himself Harry Warren was at the fore of noteworthy events: as a boy of eight, he had been scheduled to sail on the *Titanic*'s maiden west–east voyage.

As a student, Warren excelled in track and field, rugby, cricket and field hockey. At UBC in the early 1920s he obliterated the opposition while winning the 100, 220, and 440 dashes. His game-winning try, scored in the dying minutes of an intensely fought McKechnie Cup rugby match between UBC and Victoria, is still remembered for the excitement and because of who it featured: Warren and the man who took him down, Boss Johnson—later premier of BC. Three years later, while working on his doctoral dissertation on lead and zinc deposits of southwestern Europe, as a Rhodes scholar at Oxford, he won the British Games 220 in London.

One of Warren's running partners of the time was Harry Abrahams, the deer-quick sprinter featured in the film *Chariots of Fire*. And the next year, while coaching the woman's 4 x 100 relay team to a gold medal he was also a spare on the men's team at the 1928 Olympic Games in Amsterdam. There, he roomed with sprinter Percy Williams, the golden boy of the Canadian track team. "I slept in the cot next to Percy," Warren recalled years later. "His coach was concerned Percy get lots of fresh air and oxygen. But he had a nasty habit of always, before falling asleep, pulling the sheets over his head. My job was to reach over and pull the sheets off him."

When he wasn't competing, coaching or studying, Warren found time to act, and was a favourite with legendary UBC thespian Freddie Wood. Warren, dubbed "the human magpie" as a child for his non-stop chatter, was a natural on stage, his voice projecting to the farther reaches of the venue. He was also an early advocate of the League of Nations (later the United Nations).

In 1932 Warren joined the UBC faculty of Geology and Geography. The department shared his Anglophile prejudices. Among the members was M.Y. Williams, later department head, a United Empire Loyalist with a grudge. He considered that geology ended at the forty-ninth parallel, and ordered the library not to buy books on US geology.

While Warren's political views were in harmony with the department, he proved to be too eccentric to do well in the university hierarchy. As if to establish his independence, one of Warren's first acts was to deliver a speech to the assembled executives of the Canadian mining industry challenging the common practice of paying engineers in stock. The story caused a great fuss, but nothing compared to the commotion that ensued when he suggested to a gathering of business scions that patronage was holding back the province economically. The ensuing flap embroiled then BC premier Duff Pattullo and UBC president Dr. L.S. Klinck. No one ever again questioned Warren's independence.

His office was in a grey stone building where, if the window was flung open, as it often was, he could hear the shouts and whistles from the sports fields. The room quickly filled with rocks and was anchored by a large simple wooden desk into which a knife was jammed—to dismember his daily pear. The academic trappings of mortarboard hat, tassels and cape, along with metal blakies on his shoes which announced his arrival to class, became hallmarks for generations of UBC geology students. "The Robe," as he was called, taught mineralogy and petrology with a flair usually reserved for the best of Shakespeare's plays. Handing out trays of rock samples for an elementary rock identification course, he insisted that the samples be returned with the white paper exactly as found.

Warren was a small, tidy man with a dimpled chin. For five decades he was a dervish in the department, constantly reaching for ideas. The phone was rarely out of one hand, a report or rock rarely out of the other. To visit him in his office was to see intellect in action: one journalist who came away from an interview said it required "physical effort" to have a conversation with the dynamic professor. He was a radical conservative, or conservative radical, or just plain crazy.

Ted Danner had just graduated from the University of Washington when he was hired at UBC in 1954. Warren approached him, hopeful for an ally in the faculty. "He said, 'Do you play grass hockey?' I said, 'I don't know anything about it.' He snorted and walked away," recalls Danner. Though the two men would be at UBC for three decades, that was the last time Warren would have anything to do with Danner.

As a faculty member, Warren would listen patiently to a prospective master's graduate deliver an oral defence of a thesis, then ask the nervous student who the Canadian representative was on a certain United Nations committee or what the hydro rate was in Ontario. His questions got so unnerving that he was asked not to attend. At the same time he often lectured around the province. Colleagues in the department were startled to discover—often months later—that Warren had represented himself as chair of the department (he never was).

Freed by his eccentricities from the usual burdens of academic life, Warren pursued his interest in mineral/plant relationships but soon discovered that there was a lack of literature to base his investigations on and an even greater lack of enthusiasm within the university to fund such explorations, where his scoffing peers said geologists should look at the ground and not at the trees. This is where serendipity slipped into Warren's story. At that time in Vancouver there was a benevolent bookseller named William Dorbil who recalled stories he heard as a boy about a certain flower that grew only above the tin mines in Cornwall. Wondering if there could be any basis for the accounts, he granted money to the university to fund exploration, and Warren's study was launched.

Warren started by collecting samples of wood, foliage and brush from areas where copper was known to exist and from areas known to be copper-poor. Ground into "flour," the fibre was burned in a specially designed oven at 500 degrees Fahrenheit for four hours, and the residue analyzed. The procedure had to be pioneered, but Warren and his staff quickly determined they were onto something. Plant samples from around the Britannia mine, for example, yielded 583 parts copper per million, while samples from copper-poor areas yielded 14 parts per million. "We believe we have found a valuable tool to assist in prospecting, and we think this research will bring an entirely new field to light," Warren declared.

One of Warren's first studies was on *Phacelia sericea*, a purple flower with orange stamens. "The wretched flower can grow without any gold. But if there is any gold, the cyanide in its roots collects the gold and gives you a clue about what's there," he recalled. His study of *Phacelia sericea* led to the identification of three gold-bearing areas in BC. Similar work in the Highland Valley extended the known ore bodies in the area, particularly in the Bethlehem copper property. Other similar investigations followed, each stimulating more and more ideas, so that by the 1950s Warren had many projects underway. From his collections of plants on the Cariboo Highway, for example, he came to identify the presence of lead in gasoline as a pollutant. This discovery led to more work in the area of geology and human health, and his early calls—now prescient—for large-scale investigations into the relationships between trace element content of soils and rocks and uptake by foodstuffs.

Some of his most significant work took place around the giant Inco smelter in Sudbury, Ontario, and areas of industrial England but he was called upon to work around Vancouver as well. In one study he identified an area of Toronto with one hundred times more lead dust than normal. Called to investigate industrial pollution in Richmond, he discovered that horses grazing in a nearby field were dying of lead poisoning. From these and other studies he pioneered research into the startling relationships between minerals and human health. Too much lead could lead to heart and coronary disease, Parkinson's disease, multiple sclerosis and certain types of cancer. His research revealed a striking correspondence between areas with high rates of multiple sclerosis and geological areas with known high levels of lead.

One summer he collected trout livers, thinking that they might filter chemicals from streams. Reducing them in his laboratory, he cleared the building with a foul stench. In 1973 his efforts in the area of human health resulted in his installation as an honorary member of the Royal College of General Practitioners in the UK. A lifelong student of good beer, he often badgered overseas escorts into stopping at local pubs and breweries. He said this was the way he studied local trace minerals.

It was typical of Warren, however, that his enthusiasms did not displace each other, but rather accrued, much like the sedimentary rocks he studied. A keen coach in his later years, he demanded from his charges the kind of "well-played" attitude that he had been raised with. When his daughter Charlotte, a top-ranked field hockey player, let fly with relatively modest expletives on a field adjacent to one where father Harry was coaching, he sent her home. On one occasion he was reffing a game when future BC premier Gordon Campell was playing. When Campbell spit, Harry called the game and lectured young Gordon about on-field manners. Even as an old man, Warren stumped around the fields, encouraging players. "Up you go forwards! Like a pack of hounds!" His success was the undoing of his legacy; in the *Encyclopedia of British Columbia* his entry is for sports, not geology.

Within the halls of academia, one of Warren's least palatable eccentricities was his support for prospectors. Too unscientific for office-bound geologists, prospectors represented for Warren the human version of the plants that he studied: by working the land, they took up lore and wisdom that was not accessible through books. He manifested his support for prospectors by starting and supporting the BC Chamber of Mines' prospecting school, one of the longest-running in North America. And during the summers, he prospected himself. For many years his partner in the bush was Penticton native Ferdie Brent, who knew how to handle packhorses and live in the bush. Brent, Harry Warren once recalled, was the kind of man who could shoot a moose and find a use for every morsel.

Like many eccentrics, Warren was more interesting to watch than live with. His wife raised their children largely single-handedly. Family interaction was limited to watching Warren play sports on Saturday afternoon. And in his old age, a certain delusion set in. Too eccentric to ever be named department chair, he nonetheless continued to represent himself as such at meetings around the province.

At age ninety-one he was still working a claim near Big Bar. Bent and arthritic, Warren was unable to do much himself, but then he never liked digging. His son came back to camp one time to find his father face to face with a grizzly. With shades of athleticism, Warren was wielding a shovel at the bruin, yelling, "Shoo! shoo!"

In his lifetime Warren collected an enormous number of honours, including the H.H. "Spud" Huestis Award for prospecting excellence, and several presented by the Royal Family. During these ceremonies he always made sure that he pinned to his lapel the little hammer that is the geologist's insignia. Spotting this button, Prince Philip said, "I remember you—you were the little man with the hammer."

The tow truck driver at the prospecting school was exhausted from lack of sleep. His roommate snored. If Jim Simser could doze off first, he'd sleep through the racket. But his roommate was into REM before Simser was out of his boots. One night he awoke to a sound like someone dying. So he cut fingers off rubber gloves, stuffed the capsules with tissue, and rubbed Vaseline on them for earplugs. With the hybrid earplugs, he reported, he could only *feel* the snoring, not hear it.

Simser works with the BC Automobile Association so, he said, he doesn't lasso vehicles from parking arcades. "I'm a tow truck driver—don't get mad," he deadpanned to a group of students. He once towed a car from Jimi Hendrix's mother's house on Princess Street in Vancouver. Fiftyish and fine-featured, he favours black jeans, T-shirts and a worn jean jacket with the collar turned up. His hair is black and white and his moustache finely trimmed.

Simser has taken a placer mining course and was at the prospecting school to increase his understanding of hard rock geology. He has claims near the Deadman River, on an old town site. The river and the valley it drains are named for a French fur trader Charette, who was murdered in 1817 over an argument about where to set up a camp. The sides of the valley are coloured with red lava and white Mazama volcanic ash. Simser's claim is on a bench that wasn't mined. He planned to hire an excavator and trench the property. "If there is any colour I'll dig," he said. He was prepared to plumb to one hundred feet. He'd use wood cribbing until he reached bedrock. Then he'd go horizontal. The point is to get at placer gold that has settled to bedrock. From his house on Dewdney Trunk Road in Mission, it takes five hours to get to his claim "because I drive slow."

Simser thinks many modern cars are insane. This statement is not meant to be metaphorical. He says computers in some vehicles have a mind, and the mind is bad. To kill the mind, he pulls the terminals off the battery. He says the terminals have to stay off for nine minutes. Any less and the mind remains. When he told me this I heard the voice of HAL, the computer in *2001: A Space Odyssey.*

Prospecting is not an obsession with Simser but he thinks more young people should be involved. He thinks young people don't have positive heroes; music stars and sports heroes don't set a good example. Prospecting, he thinks, encourages independence and enterprise. The Deadman River, as it happens, attracted another alternative-thinking prospector. In the early 1990s a Tibetan monk arrived in the valley, following his master's instructions to find the Centre of the Universe, which the monks called the core of existence. The monk stayed at a privately owned lodge and one day, while wandering the property, came upon a bluff. A tourist brochure explains the rest: "Performing his Buddhist rituals to investigate all eight directions of the compass as well as the elements existing in the eight realms beyond, the mysterious visitor eventually reached a state of mind in which he received the revelation that his quest had been completed. He had indeed, so he excitedly informed the lodge owners, found the Centre of the Universe!" People who have camped on the land since then have reported many strange occurrences, including a mysterious sound described as "the Mormon choir, but without words."

I was sprawled on my bunk when Don Conley marched into my room, adjusting the fly on his trousers. He's ovoid, dressed in work pants, button-up T-shirt and suspenders. His face is shadowed in whiskers like Fred Flintstone and he holds his mouth in a permanent pout. He looks like a classic line screw.

"Hey, writer boy!" he said, still adjusting his pants.

I reached for my geology hammer. He wanted to know where he could get a book, a local history called *And So That's How It Happened*, by W.M. Hong. It is about the Barkerville area, where Conley has five claims with a partner. He said that a seventeen-ounce nugget came off the property in the 1920s. He thinks there are more. Last year, using pick and shovel, he and a partner fetched six ounces. Their claims are in a V-shaped valley. The creek at the bottom is subject to many environmental regulations. Conley can't run equipment on his claim, which he'd like to do. He made a pistol with his index finger and thumb and pointed it at his head. "The pulp mills can put chemicals in the river, and a million turds come out of Vancouver, but can we run an excavator on the property? Oh no."

His secret weapon, he thinks, may be his partner, who is apparently gifted with the attributes of Einstein *and* Newton. "He has an MA in Education. He's in Mensa. You know that movie, *A River Runs Through It*? He'll watch that with an encyclopedia on his lap. He'll say: 'Don? You know why an elephant can't jump?' And I'll say, 'No, I don't know why an elephant can't jump.' And he'll say, 'Because it has four knees.' That's right."

He stretched, made a comment about his sciatica and then, apparently thinking he'd gone too far in making much of his partner and not enough of himself, said, "You talk about writing. When I went back to school after the army I had to write an essay. I'd grown up on a farm and knew about animals and all that shit. Assiniboine College wanted two pages. I handed in forty-two. The subject was 'The Effect of Tuberculosis and Bursitis on Wood Buffalo and the Environment.' No one at Assiniboine could understand it. So they sent it to the veterinary college at the University of Saskatchewan. They passed it, but not before taking a copy for their files."

He stormed out and I heard him crash-banging next door. In a moment he returned and opened a big fist in front of my face. There were a dozen or more multicoloured, multi-sized pills. "Do you know what you do with these?" he asked. His shirt buttons, under the strain of an impressive jailer's belly, were bull's-eyed at my head. He tossed the handful in his mouth with a dry clinking sound and chased them with water. I realized he was a variety show, and if you don't like one act be patient because another will soon appear.

My roommate, Dan Wilson, came to the door. There was no way the three of us could co-exist in the room. Conley went out and Wilson stepped in. I thought Ed Sullivan should say good night. There was a brief flurry of activity, then the bunkhouse emptied of noise as rapidly as an elementary school after the last bell. Wilson clicked his nightlight off, then I turned mine off. I saw a fractured red wall behind my eyes and went to sleep.

Day four. The morning air was as warm and thick as a sweater. The evening before had been chill enough to warrant a wool hat, now it was flip-flops and shorts. Birdsong was audible at 4:30 a.m.; by 6 a.m. it was such a racket you'd have to be in a coma to sleep through it. Unmuffled two-cycle engines would have been more blissful. Patrick Monk and several others were already up. The gravel between the buildings had taken on the dryness of talc, so that to walk carefully was to stir up a veil of dust. Our feet were filthy.

Perhaps sensing a sort of wooden-bench-inspired insurrection, Stevens started the day with a promise that this would be our last major classroom session. It didn't do much good: as we sat down, Kerry swallowed a yawn and made it look as though it were sour. The lecture topic: exploration techniques. "Say you have discovered a potentially significant mineral showing through your prospecting efforts," said Stevens, stabbing a finger at the relentless PowerPoint. "Now it's time to advance the showing before you try to sell your property to a company. What do you do?"

The first, of course, is mapping. Maybe it's some lingering adolescent reaction to teacher-repetition, but I was fed up with mapping. It prefixes every geological conversation, just as gardening advice is always prefixed with, "Your plot should be weed-free and well-drained . . ." Really? I think the point is that there is no use doing extensive, detailed (and often pricey) prospecting if you haven't mapped. I relented: hockey players work within the confines of the rink, swimmers within the limits of water. Prospectors work within the limits of maps, and that allows them to use the most common of exploration techniques such as geochemistry.

Geochemistry is one of those five-dollar words by which first-year geology students elevate themselves above their non-geology counterparts. The wider purpose of geochemistry is to narrow a search. The search may be regional, local, or within a property. Since the days of nuggets are largely gone, geochemistry has provided the impetus to bonanza.

During the 1970s many provinces, including BC, conducted massive geochemical studies. This information was of the sort that every exploration company wanted, but no individual company could afford. The release of this data was of such importance that it warranted D-Day-style embargos. Copies of the studies were shipped under the tightest security to regional government offices. Release was scheduled for 9 a.m. on a certain morning. The night before, prospectors and field crews for exploration companies stayed in the same hotel. Instead of all-night piss-ups, the halls were quiet by 9 p.m.

In Terrace, the study data was released at the airport. Representatives of nine companies tore open their copies. Some—like Muslims at prayer—laid the maps out on the tarmac; others trusted their ability to decipher the studies and sprinted for the dozen helicopters that sat, blades whirling, on the runway. Under a ceiling of evil-looking cloud, like a hatch of flying ants, the helicopters rose simultaneously and vanished into the clouds.

Stevens handed me a shovel, indicating that for now, at least, this was as sophisticated as our geochemistry was going to get. We were on the Fairview Property, where the prospecting school's final assignment took place. It was 2 p.m. He pointed at the ground and said "Dig."

Full tuition and I was going to dig? On my knees, I stabbed the short-handled shovel at the soil. It was moist, powdery. "Clayey-sand," Stevens announced. I dug through the humus and into mineral soil, what geologists call the B horizon. Rainwater leeches minerals into this layer. By analyzing the sample, a lab can detail which metals are in the soil, and in what quantities. It is like sniffing the dirt. It was remarkably easy and Stevens assured us that if we were standing on a cliff, it wouldn't be. I palmed the yellow soil into the bag and my partner, Richard Barkwill, marked it. It was as easy as taking out a library book.

With enough of these samples, a geochemist can guide his or her research. It is something like the game Battleship: hit and miss until you get the sense of where and in what direction an anomaly lies. On a river with tributary streams, for example, geochemistry may indicate anomalous amounts of gold up to but not beyond the interception of a creek. In such a case the geochemist would target further investigation up the secondary stream, and so on.

On paper, geochemical exploration looks very sound. In the field, however, there are many forces that disrupt the science. In much of Canada repeated glaciations have bulldozed and smeared the overburden, spreading evidence far and wide. And human activity is having large-scale effects as well. A massive geochemistry project outside of Flin Flon, for example, had gone amiss because it failed to take into account trace mineral fallout from the city's giant smelter upwind.

Shovel in hand, Stevens set off to find the next group for a soil-sampling tutorial. There was an hour remaining before the bus left to take us back to camp, so Barkwill and I decided to stroll the property. The urbane Drew, our other partner on this project, was feeling faint and dozed in the shade of a tree. The property resembled an overturned ship, keel in the air. It sloped away sharply to the south, where the grass was already an off-green. By July it would be straw yellow. The grass was broken only by outcrop. We noticed freshly broken rock, the work of other students. Walking clockwise around the property, we entered the cool shade of pine forest. The transition was stunningly sharp, as if an oven door has been suddenly closed.

We were supposed to be scouting rocks but we got sidelined investigating a rickety bicycle jump. Apparently the ambitions of the neighbourhood boys have leaped property bounds. They had installed all sorts of jumps and runs, some of which looked improbable. One run was a walkway of flimsy one-by-fours set on piles of mud and ending above a draw. Evidently good sense prevailed, because the whole thing looked unused.

We walked past several rock piles before recognizing them

for what they were: pit excavations. Unholstering my rock hammer, I broke a piece and studied it under the loupe. Pyrite. Barkwill and I were both at the stage where we recognized that it should mean something, but we didn't know what. On the north side of the property, there were many signs of old workings: a roadway, more pits, and several massive trenches. Across a draw and beyond the property limits of the exercise were the remains of a track, where the waste rock was dumped. We finished the day by walking the lush and easy north side, the kind of meadow that seemed more suited to farming than mining and arrived back at the truck, where an enormous pounding of hammers was going on.

What we saw when we broached the hill was, perhaps, as close a re-enactment of ancient mining techniques as is likely to be seen. A dozen men wielding hammers, breaking rocks in the hot sun. Stevens, apparently, had suggested that they win some ore from the rock piles, which we'd later smelt over an open fire. I mentally deleted the sunglasses and vests and the scene seemed positively Egyptian-era, when, according to one historian, a certain class of worker lived and worked underground. The separating of ore from rock was physical and seasonal. Evidence suggests ancient man knew where to find rocks that suited his need, just as we know what aisle to seek food in the grocery store. They took what they needed and no more.

After dinner Don, a bemustachioed and resourceful carpenter from Logan Lake, found a pipe and with a metal cutting tip fashioned it into a pounder. Dave Watson produced a metal gold pan, and several students took turns pulverizing the rock. It grew lighter in colour the more it was broken down. There was a bitter taste in the air. Meanwhile, another student, Randy, set about dragging branches from the nearby woods and lit a fire. There was the spirit of industry in the air, and those who weren't working chose to stand instead of sit. The pan went on the fire.

Sometime during the day, we'd all had the same collective thought: we realized we were not students in the traditional sense and that we could all tell Stevens to go to hell if we so chose, without consequence. We all fetched beer from our rooms with the full intention of getting drunk. I know I did. Stewart, a lawyer from Vancouver, collected money for a liquor-run to Oliver. Most of us hadn't even touched our supplies, but nonetheless we ordered more. The spirit of excess was upon the camp.

Kmett, the American, was proud of his new Toyota ("I don't owe a thing on it" I heard him say, repeatedly); he drove it up to the fire, opened the door, and stood, one sandalled foot in, the other tapping on the dusty ground. He blasted country and western. The sun was low and seemed to be tumbling on a steep incline. The coming of dark made us heap the fire even higher. Don Conley produced—from nowhere and for no fathomable reason—his passport photo. Everyone had to look. "Handsome, eh?" he asked.

A leggy woman walked out from a nearby ranch house, stretched languorously, and strolled across the field. For all the hormonal moaning we might have been twenty-year-old sailors arriving from two months at sea, instead of forty- and fifty-somethings rubbing joint cream into our ankles. The fire was hot, but not hot enough to smelt ore. Randy discovered a wall fan and someone else found a piece of sheet metal. After a lot of arguing and a lacerated finger, the sheet metal was bent into a shroud with the fire at one end and the fan at the other. The flames turned to amber coal. To even look at it hurt the eyes.

Every hour or so, Dave lifted the top lid off with barbecue flippers. The volume of ore was getting smaller and smaller. Gary turned off the country and western music (thank Christ!) and went to sleep in his truck. At midnight I passed around a bottle of scotch.

Day five. At breakfast one of the cooks said, "Snakes will be out today. Better not reach in any holes." He's low-slung, with a shaved head, eyes wide-set behind large spectacles. I got the feeling that both cooks snicker at our enterprise, as if we were studying to grow organic radishes. Baldy used to prospect around Grand Forks. He says, "I must have hauled thousands of pounds of pyrite out on my back. The geologist would look at it and say, 'It's just iron.'"

The sun was a fierce disc by 8 a.m. It was seventy-five degrees Fahrenheit. Kmett wore a heavy pullover and complained of the cold. We were joyous in the at-least-we're-not-being-hit-on-the-head-with-a-brick way: No school! No wooden benches! No sore asses! By the firepit's ashes, Don, the carpenter, panned the roasted mixture. He placed a handful of material in the plastic pan and then tipped the pan into water, as if it were sipping by its lower lip. Using all wrist action, he got the water rotating in the pan and every few seconds tilted it so the larger particles spilled out.

Though I was blurry, the word-root action came on. Prospecting: NHL scouts go prospecting for junior talent; something doesn't pan out means it didn't work; we talk about the nugget, or fool's gold; recently the BC railroad shut down its half-century-old passenger service. It was called the Cariboo Prospector.

Don reduced the amount of material in the pan to a vialful.

He swirled the water so gently that I wondered how he could make his wrist move like that. There was very little black sand, or magnetite, which is not a good sign. He stopped panning, pointed and said, "There." I leaned close and squinted. I saw tiny trails of orange sparkles. So that's what is meant by lustrene—a dissatisfying wimp of a word that means a suggestion of something. He showed the pan around, and then tipped it on the ground.

IV North Again

Though bush pilot Dick Boen has been dead for some years, people in the little town of Atlin, BC, still talk of him as though he was lounging with his boots on his paper-strewn desk, fingering a cigarette, as he did for many years when he wasn't in the air. Boen was slight—just 115 pounds—and an excellent pilot. He could nurse an overloaded Beaver off a lake and make it soar like a kite. When it came to customer relations, however, he was a thumbless calligrapher—pretty much useless. He hated tourists. Once, when freighting hunters and supplies into a remote lodge, a passenger reached for a can of beer. "Not in my plane you don't," snapped Boen, while at the same time lighting a cigarette. "What about the 45-gallon drum of fuel?" said the passenger. Boen would have none of his logic. "If it goes it will be quick!" he snapped.

Chris Moser has few of Boen's qualities—except he is a well-regarded pilot. For that I was most grateful. I was one of two passengers—geologist Dave Caulfield was in the back, nestled against a 45-gallon drum of fuel—on a heavily laden Beaver that Moser was coaxing off the glassy surface of Atlin Lake on an early October day. We were bound for the Thorn, a much-studied, puzzling property in the remotest corner of northwest BC.

Big and beefy, with thinning hair and a pleasant face, Moser pulled at the controls like a person might signal a car backing up. His hands were etched with engine oil. Each time he pulled back, the plane seemed to try and break free of the lake. I looked out the window. Beyond the froth of the pontoon the lake reflected the hulk of a mountain opposite the community. The sky was a depthless grey of cloud. Finally, with the gentlest coaxing from Moser, the aircraft lumbered into the air, ascending more on power than aeronautical finesse. Below us was the alluvial fan of a creek, site of old gold workings and a half-suburb where leashed dogs had run crop circles into the ground, and little sawmills stood, dust piles looking like the leavings from powder bugs.

Only when the aircraft had climbed over a thousand feet above the lake's surface did Moser relax. His shoulders set back slightly and, for the first time since we got in the plane, he spoke. His words, heard through a headset clamped to my ears, clanged far inside my head. "The Beaver doesn't want to turn when it's climbing. It just kind of falls over." He's been flying for years—in Red Lake, Ontario; Edmonton; Bella Coola. He got sick of the flying business once and got into construction. He was working with sheet metal on a project in Edmonton when he heard the distinct growl. He looked up and saw a Beaver and simultaneously dropped his tool belt. "I said, 'Screw this shit. I'm out of here.'" He leased the Beaver and calls northwestern BC his workspace.

Though geologists like to tell stories of pilot derring-do, they much prefer to fly with dull, competent pilots. I thought of a conversation I had with Matthew Kempthorne, a young man who spent five years flying around the north on mineral exploration projects. "Pilots are always under scrutiny because everyone's life is in their hands," Kempthorne said. "One time I was flying back to Whitehorse from a camp and the low-fuel alert came on. It went 'Ding!' One pilot looked at the other pilot and said, 'Pizza's ready.'"

An outfit Kempthorne worked with had a pilot dubbed Freddie the Drag Queen. "He set us down in a remote area and shut the machine down. After a while I noticed something moving on the rocks. It was the pilot, naked, flitting about." Against the likes of Freddie, Kempthorne contrasted a helicopter pilot who came on the radio during a flight and calmly said, "Boys, I've got a problem. Got to set her down." It turned out the problem wasn't little at all—a compressor had failed and the engine was about to stop. But he remained calm and set the machine down safely.

Our destination was a kidney-shaped body of water that appeared as a dot on all but the most local of maps. It was called Little Trapper Lake. The area is so remote that it does not have a regional name, like, for instance, the Iskut, or the Spatsizi. It is scribed, roughly by the arc of the Taku River on the north side and the crab-like extension of the Sutlahine on the other side. It is the size of Belgium yet it is nameless. The landscape is voluminous and varied rather than grand, like the Rockies. Fifteen minutes into the flight the craft passed over a lodge at the confluence of the Taku and another river. With the exception of a cabin at Little Trapper Lake, there was nothing to mark civilization until Juneau, Alaska.

After fifty minutes a small kidney-shaped lake came into view. Moser brought the plane down low over the treetops and promptly set it on the lake—not a postage-stamp landing, but all the same a landing that suggested a man keen on getting the aircraft onto the water as soon as possible. Engine idling in a pots-and-pans-in-a-drawer clatter, the plane taxied to the side of a homemade dock about the size of a good pool table. A cabin peeked through the trees. An immense, surly-looking man, clad top to bottom in denim, lurched to the dock, took the rope and lashed it to a cleat.

Caulfield stepped out—unfolded, really, for he's over six-foot-six—then me, and with no more than a grunted greeting we formed a bucket brigade passing packs, groceries and goods hand-to-hand, from the plane, the wharf and onto the beach. Last to be unloaded were the fuel drums. We tipped each barrel on its side and down two metal ramps leading to the dock. I grabbed my bag. The big man said, "Dinner is at five." Chris Moser was already untying. He looked at the sky. "I want to get home before dark." He stepped onto the pontoon, and shut the door.

An intriguing prospect, remote and with great potential, the Thorn property not only embodies many of the qualities and challenges facing exploration geologists, but it does so two- and three-fold. Where many prospects host one kind of mineralization the Thorn hosts three. The Thorn—260 kilometres south of Whitehorse and 750 kilometres north of Vancouver—could not be more remote and inhospitable country if it was plotted. Its history, spanning three generations of geologists and a revolution in geology, enfolds the history of exploration in BC, and includes tragedy, persistence, disappointment and science.

Until the 1950s, most of the ore being mined throughout the world was found because it outcropped. A prospector, pick in hand, eyes to the ground, wandered the bush breaking rocks. When he broke a strangely coloured rock, more rigorous exploration began. If reserves-per proved up, a mine opened. This was such a system that the histories of the world's great mines always included in their first pages an account of a man in dungarees and a floppy hat, often yanking a recalcitrant burro. By the 1950s the yield of new discoveries from this old method was fast diminishing. Economists and resource managers reported to the president of the United States and other world leaders that the world would soon run out of copper, lead, zinc and uranium. The Malthusian argument prophesied economic disaster for lack of minerals to keep the industry turning.

Among the few who vehemently did not believe in the portents of gloom was a geologist in Australia named John Sullivan. In the northern area of that country he combined aerial photography, elementary ground physics and geochemistry to try and understand the environments in which ore might be found. Impressed with Sullivan's work, mining giant Kennecott Copper Corporation invited him to come to Canada and help its exploration offshoot, Kennco Explorations, apply his techniques to Canada. As with many other companies, Kennecott's mines had been located by prospectors. But the company was mining 30 to 50 million tons a year from deposits discovered fifty to one hundred years before. It didn't require a degree in supply management to know that new reserves had to be located sooner rather than later. In an interview, Sullivan recalled his impressions: "My first impression from a visit in 1952 and from initial inspections in 1954 was of the extreme difficulty of coping with [Canada] using only human hands and feet—vast forests where you could practically see nothing but trees, tremendous mountains, rivers, lakes and huge areas covered by glacial debris."

Such conditions, thought Sullivan, lent themselves to geochemical techniques, and he brought in Herbert Hawkes to educate the company's geologists in the use of geochemical techniques. Success was immediate. In New Brunswick the company located a two-million-ton, 2-percent copper deposit, and in New York State they found the Murray deposits, totalling twenty million tons of lead-zinc-copper ore. In BC, the emphasis was on porphyry deposits. Sullivan had worked in the Andes and thought that the best efforts would be in looking in the areas immediately east of the Coast Mountain granites. The company invested in laboratories and large-scale geochemical surveys. When they identified an area they flew in a prospector who, in turn, found float and tracked it to the source. Within a few years Kennco's discoveries in BC included deposits estimated to contain five million tons of copper and one billion pounds of molybdenum sulphide.

Kennco's discoveries were soon followed by discoveries of other companies, including mining giant Anaconda, which was one of the first companies to identify the Thorn's mineral potential. Anaconda, like Kennco, was after porphyry copper deposits. They identified a rich veining system at the Thorn but didn't pursue it because it didn't fit the model of what they were looking for.

Vancouver's Julian Mining Company worked the Thorn property in the 1960s. One of the crew was a short, agile young geologist named Bob Adamson. Adamson has very likely had more experience in the Thorn than any other person. In fact, he named many of the area's central features—he called the property "the Thorn" for the devil's club that grows so rankly on the mountainside. And La Jaune Creek—which Adamson adopted from a French-Canadian prospector's mangled pronunciation of yellow in French.

According to Adamson—now in his seventies but with an amazingly precise memory—the Thorn has never had its true potential understood because companies insist on trying to make it conform to a certain kind of mineral deposit when in fact it is of another sort altogether. Adamson believes there is potential at the Thorn for a successful high-grade underground mine. He argued this regularly at the Thorn with Anaconda's geologist, Glenn Waterman. Behind Julian was the mining giant Anaconda. In its day Anaconda held sway over the political fortunes of South American countries. In the 1960s Anaconda's top geologist was set on finding more porphyry deposits—which could be transformed into big tonnage, open-pit mines, and which had made the company very successful. A number of factors at the Thorn didn't work to that end, but foremost is the fact that it can be almost impossible to change the mind of a corporate geologist with a dose of hubris.

Adamson's efforts to convince Anaconda to look at the Thorn property in a whole new way would likely have failed anyway—but a tragedy effectively gutted the company's interest in pursuing the area. On September 21, 1964, a Beaver float plane flying the colours of Trans Provincial Airways set off from Trapper Lake with three prospectors and the pilot. It was routine take-off—until the aircraft suddenly turned on a wing and spun into the lake where, laden with drill steel, it quickly sank.

The lost plane was eventually located in 230 feet of water. A salvage crew flew in, constructed a raft and winch to haul up the wreck. The job was made miserable by bitter winds and plunging temperatures. Winter was fast approaching. If the plane wasn't pulled from the lake quickly, it—and the dead aboard—would be entombed in ice until the next spring.

After ten days the Beaver had been moved just thirty feet up and dragged closer to shore. Then a second, smaller disaster struck. While the salvagers were working on the lake their camp burned to the ground. Exploding ammunition kept them at a distance. The fire destroyed almost all their gear, including the radio transmitter. It was a full two days before officials in Atlin sent a plane to investigate. Flying over the snow-covered lakeshore, the pilot, flying a wheel-equipped plane, spotted a message trampled in the snow: "Burned out." He tossed food supplies before returning to Atlin.

Efforts now focused on rescuing the salvagers. There really was a classic race against the coming winter as the lake was about to ice over—making it impossible for a float plane to set down safely. Once the ice had begun to form it could take weeks to form a surface strong enough to support an aircraft on skis. With temperatures dipping to freezing and below, the wings of another Beaver were slathered in grease to help prevent icing. A pilot named Herman Pederson flew in, set the plane down on the surface of the lake and collected the cold, hungry salvage crew. The wreck and the dead were left for the winter. The bodies were not recovered until spring of 1965. Soon Julian Mining gave up the claims. Recalls Adamson, "Psychologically we were out to the Thorn."

Since the 1970s a number of companies have tried to solve the riddle of the Thorn. Most recently, a small Vancouver junior exploration company had committed about three hundred thousand dollars to a drilling program that, they hoped, would produce evidence to interest a bigger mining company to invest in yet more exploration.

The gas light in Brian Mercer's cabin hissed like a strong wind. A honey glow spilled onto the wood floor and walls, onto bookcases and the woodstove. The stove was ingeniously created out of a forty-five-gallon drum, and pumping out heat at such a rate that the coveted spot at the kitchen table was by the window, where a mountain-kissed, water-cooled breeze sent the white curtains dancing. Dinner was chicken cacciatore, green beans, mashed potatoes, meat loaf. Two kinds of salad. It was a camp, but the table had the feeling of a farm kitchen at lunch.

Including Bill and Lloyd—the two drillers up on the mountain—and myself, there were twelve of us in the camp. They included the two day-shift drillers; Henry Awmack, the geologist whom I'd met before; Norm Graham, the helicopter pilot; Tim Sullivan, a prospector; the cook; the cook's helper; Brian Mercer; and Awmack's business partner Dave Caulfield. One of the drillers, a big man named Dan Bomford, used to date a woman who worked for a cosmetician. He said, "The bad news is that men go in and get shots in the calves, the chest. They get shots in the butt."

The next day I went to work with the drillers. "Drillers go first" is the motto, so while Caulfield pored over maps and got his gear ready, I ran through the woods to the chopper landing. A neophyte at riding in choppers, I always tried not to be the last one in, so that responsibility for shutting doors was not mine.

After a ten-minute ride in the muddy dawn, the pilot set us down on a gravel bar on La Jaune Creek. The creek was narrow enough to throw a stone across, but not so narrow or shallow that you would want to ford it without striking a fire on the other side. We greeted the night-shift driller and his helper—two of the dirtiest, most raggedly dressed men I have ever seen. They said there was a problem last night. A pump had thrown a piston. No pump meant no water, no water meant no drilling. We bent to study the block. It was cracked.

While the drilling crew discussed what to do, I studied the surroundings. Even in the early light it was easy to see what had attracted exploration: the gossan glowed like wheat. It felt like it was throbbing. Wooden pegs flagged with survey ribbon marked the path the drill would take deep underground. Routing the path of a drill is a difficult decision. The dimensions of the earth are never so large to a geologist as when he is planning a drill hole. Straight down? Forty-five degrees?

Furthermore, every geologist is aware of stories about drilling that, for whatever reason, did not go as planned yet returned amazing results. At the site of what was to become the Craigmont Mine, for example, a geologist left a night-shift drill crew with instructions to drill until he told them to stop. He left for a local bar, where he stayed far too long. The drill worked far below what the geologist had planned, but turned up results that helped confirm the ore body. And near Flin Flon years ago a geologist ignored his boss' instructions in favour of his own hunch and drilled what became the legendary Forbidden Hole, which identified a major ore body. The geologist responsible for the find personally carried a heavy piece of the drill core into the plush office for his boss, who was sitting behind a polished mahogany desk. The geologist held the core over the desk. "Support for my fucking theory," he said. Then he let the core drop.

We hiked up a trail crossing the gossan, winding our way through tree roots until we reached a wooden platform about the size of a suburban TV room. It was partially dug into the hillside and partially overhanging the bank. As at Mercer's, I had the feeling of a land of two stories. There was a summer story and a winter story. Some years before an avalanche had gone through just ten paces to the west here, clearing a swath 150 metres wide. But the fierce winds that accompany avalanches had snapped off the tree trunks on either side, tapering the avalanches' effect much like a haircut.

The platform was outfitted like a factory floor. A driller is, essentially, a factory worker whose workplace is in the wilderness. Within the twelve-by-twelve square there was a generator, hydraulic pump, toolbox, water tank, drills and associated parts. All had been lifted in by helicopter. Part of the bank had been mattocked out to create a space for a tool chest. At the high side of the platform there was a tool rack made of two-by-fours with nails pounded in. The two-by-fours were pine. They were made in Prince George. Draped around the outside were lights on a string, giving the platform a carnival look. Everywhere there were signs of improvisation: things like a branch duct-taped to hold up a length of electrical wire.

Dan Bomford took a hose from a grimy plywood box, hoisted it on his shoulders, and struck out across the avalanche chute. The route was uneven, rank with buckbrush and devil's club. I took another hose and followed. A big man, Bomford looked like he could lift a railway flat car. But like a lot of big men he doesn't move through the bush well. I, who weighed about a hundred pounds less, could get away with stepping on smaller limbs. Across the chute, Ian, Bomford's helper, was on his knees beside a vigorous little creek, making a dam of boulders. He'd anchor the end of the hose in a filter, and gravity would send it across the slope to the drill.

We got the hose running into the tank, and Bomford started the drill only to shut it off moments later. There was something wrong with the controls.

Writers generally thrive on bad news and other people's ill fortune, but I felt uncomfortable. For ten minutes Bomford stared at the platform. Then, moving slowly, he took a monkey wrench the size of a baseball bat, undid two hoses and traded them around. The night shift had the hoses hooked up wrong. He started the drill, pushed a casing into the ground, then sent the drill down. Ian had laid out a core box. This was to be the routine for the rest of the day.

Like a factory or a truck, a drill must work non-stop to be efficient. The driller and the driller's helper are tied to the machine. The sense of activity imported by the noise of the engine pervades the work deck. When Ian was not pulling core he busied himself—picking up juice boxes that Bomford had discarded by the dozen, tacking a nail in the tool rack. He spotted a five-gallon plastic pail, emptied it and scrubbed it out with water. It makes a tool container. The default activity of a driller's assistant is cutting burlap. Drillers go through burlap like a pastry chef goes through flour.

The driller's controls are water and push. The drill normally pushes at two hundred pounds. In sandy rock a driller can turn down the water and push harder, packing material into the casing rather than washing it away. Somehow, though, the drill needs a form of lubricant. It can be greased, but the sand sticks to grease. Typically you don't use grease with sand. There are agents you can add.

Drillers make a lot of money. They are paid an hourly rate

with production bonus. A driller can make ten thousand dollars per month. This is more than most geologists make. Bomford knows a driller in Smithers who retired at age twenty-seven, two years over his goal. He started as a driller's helper at age seventeen and worked three-hundred-plus days a year thereafter. He drank beer freely but didn't piss his money away. He bought a house, a good truck and an ATV, and socked away enough money to ensure a generous annual income.

With his hand on the controls adjusting pressure, Bomford had the preoccupied look of a man reaching his hand far into an air vent. Much information is returned to the driller via the sound and feel of the drill. If a drill goes into sand, the sand absorbs water from the drill. If the water rate is not slowed or shut off, the sand absorbs water to the point where it gets pressurized and squeezes. This pressure clamps on the drill tube. When it gets great enough it causes the threads of the tubes to squeeze together. The squeezing forces the tube to expand. Belling, as it is called, can occur very quickly and ruin drill pipe. Pipes cost seventy dollars each.

By mid-morning the drillers were into a pace. Water was bursting from the creek-fed pipe at such a rate that Bomford turned a tap off. He windmilled his arm. Ian tapped core from the overshot into the wooden tray. His job is 90 percent clasping.

Bomford, a prop with the Castaway Wanderers rugby club in Victoria, won three BC titles, most recently in a brutal match against Port Coquitlam. Long ago he damaged his shoulder and now he windmills it. When he was driller's helper his arms went numb. Asleep in bed after work, he went to roll over. His arms wouldn't support his weight. He fell out of bed. Handling cold steel all day leads to problems. Carpal tunnel syndrome is a common affliction, "just like typists get," Bomford says. He's worked for Falcon in Papua New Guinea, Argentina, Bolivia. In Peru he heard a story about a driller who did 550 feet in a single shift. Bomford's best is something over 250 feet.

Thirty feet. The core was broken. Ian battled to get it out. The drill entered the ground parallel to the slope, so the ground was broken. In Papua New Guinea the ground isn't hard anywhere, the land is too new, it is still cooking. The core comes out in particles.

At six o'clock Bomford pulled the last core and idled the engine. Ian had already organized their gear, so they simply walked away from the platform. We slipped and slid down the bank and waited at the gravel bar for the chopper.

Bomford looked at the bank. That explained what was happening with the drill. It worked hard going through sand, so it sounded like it should be going through rock. But when they pulled the rod, there were two busted rocks that looked too small to fit on Elizabeth Taylor's finger.

What happened? The water washed the sand away. The drill bit could push loose rocks aside. Bomford said, "Lots of work but nothing to show for it."

I passed the next morning in the cabin, lolling, lazing about, eating like a workingman yet lounging like a tourist. Brian Mercer was going out on the lake later and asked if I wanted to join him. Bill and Lloyd, the two evening-shift drillers, arrived at the kitchen loose-limbed and cheerful. Bill, the helper, stumped through the living room and peered in a mirror. His slim face was scrubbed except for a daub of black grease that escaped the shower. He laughed at himself and wiped it away with his shirt collar. There was an underlying smell of hydraulic oil and soap.

Lloyd rubbed his hands over his stubbled face. He had

blue eyes and a broad face, framed by a well-tended dirty-blonde beard. He looked aesthetically Scottish. He'd been on night shift for four years, most recently in Cusco, Peru. When he's finished at the Thorn he's off to Vietnam. He was raised in the Codroy Valley, on Newfoundland's southwest coast and studied at the Fisher Institute in Cornerbrook before going to study wildlife ecology in the field at Kluane Park. "I stopped in Smithers and now I can't get out," he lamented. Looking for work, he met drilling company owner Tom Britton. Tom said, "Have you ever worked in drilling?" and Lloyd said, "No, and I won't unless you hire me."

Lloyd has worked for Major, JT Thomas and most recently for Kluane International Drilling. For a long time he was in the Arctic, drilling for diamonds at Snap Lake. His favourite place was Peru, where the driller's helpers got twenty dollars a day and felt filthy rich. He has an open wound the size of a banana slug on the back of his arm.

Lloyd walked out and Norm Graham came in. Graham flies in the morning and afternoon. Midday he works on the machine and rests. He's tall, lean, business-like. There is none of the flyboy strut about him—of the sort that would, say, hoist an old Coke machine onto a mountain peak just to piss off the hiking crowd, as one pilot did in the Yukon. Graham is cautious and studied.

Graham used to run five helicopters and spend most of his time behind a desk. But a combination of reduced expenditures and difficulty getting pilots to stay in Atlin forced him to reduce his fleet to two. He flies over 450 hours a year now.

He fell in love with flying as a kid. He started out sweeping floors at Northern Mountain Helicopters in Revelstoke, then got his B licence, which permitted him to work on the dynamic components of a helicopter—the transmission, squash plate, the mast—what he calls "the moving parts." Meanwhile his older brother Larry, a pilot, was regaling him with stories about flying geologists around in the High Arctic, or teaching the Chinese to fly big Sikorskys. When Northern Mountain broke up Graham followed one of the partners, who formed Capital Helicopters. He trained one hundred hours on a Bell 47 and four hours on a Jet Ranger. With those kind of hours, he said, "you can't really fly a helicopter. You are a hazard to yourself." One of his first jobs was to fly support for a veteran pilot fighting a massive forest fire in Alberta. Graham met the pilot at the airport—the pilot had just quit. In the next two weeks Graham doubled his total flying hours.

After wrestling all morning to wire a new electrical receptacle onto one of his Honda generators, Brian Mercer was ready to go. The slowly methodical work habits of bushmen are often interpreted as an indication of excess time. But there is a reason rooted in survival. To slip into the hurry mode is to risk your life when working alone.

His skiff is fourteen feet long and planes easily. He stood in the back, not a lifejacket to be seen for miles. I felt feeble for even thinking of lifejackets. Trapping is Mercer's consistent income, augmented by fisheries work. Exploration income is something new. Everything he shows me would have to be reckoned with, should the Thorn apply to become a mine.

The salmon enter the Taku River in early June and swim steadily until they enter Little Trapper Lake, then make their way to a river on the north side that leads to Trapper Lake. I asked Mercer what the river is called and he said, "You can call it Trapper Creek." We powered up the river, swerving from side to side. He read the currents for shoal and boulders. The ride was three-dimensional, as if we were banking in an aircraft. Branches whipped us in the face. Sal, the Border collie, just blinked as the branches lashed at his face.

Around a bend the river widened and the water ran fast over boulders. "That's as far as we go," said Mercer, and cranked the boat around. To me the river looked unchanged, but he saw otherwise. We powered downstream and hauled out on a gravel bar near the lake. Scattered about were the skeletal remains of jaws and spines, milky white against the river rock. The lightest layer of protein slime covered the underwater rocks. The rock structure of the riverbed is critical to salmon habitat, said Mercer. The eggs lodge between the larger rocks, which protect them during their most vulnerable stage, when they "eye up." At that point they are vulnerable to shock of temperature or chemical changes in the water. After that, he said, "you could play marbles with them and they'd be all right."

In a cottonwood a stone's throw from us, he's built a platform. A keen student of the local bear population, he'll run his skiff up the centre of the river in the height of spawning season and then head it straight for a tree. He'll tie up and scramble up the tree to his platform without touching ground. The idea, he said, was to keep his scent as far from the bears as possible. The bears know he is there, but he is there in a non-threatening way. "To a grizzly, scent is like vision is to us. The closer the smell, the more detail." In a single season he's counted at least sixteen different grizzlies at the river, and in all likelihood there are many more indistinguishable bears as well.

We headed across the lake, the skiff pounding off Codroy-Valley-like waters—whump, whump, whump. We appeared headed for an unremarkable bay in the lake, fringed with green and yellow brush and capped with overhanging pines. It was only when the skiff increased speed did I realize that we were in a current, where the Kowatua River sucks at the lake. It felt as if we were being pulled into a throat or bottle. The power of the river made us think of being on a conveyor, and that the dinky efforts of the little engine would never counter the force.

Incongruously, a steel fence appeared in the river. The posts were anchors for a weir Mercer erects each summer to guide and count the returning salmon. A piece of white plywood anchored to the river bottom helps him see the passing fish. Mercer and his son have been tallying fish for two decades. They used to carry the supports, weighing some two hundred pounds, out in the river, but they've developed a way of tipping them from the boat. With the water this high and running this fast, he said, he'll have trouble getting them out this year.

He nosed the skiff onto a sandy beach and tied the painter to a willow. A trail follows the river downstream, skirting tree roots and occasionally edging over areas the river has undermined. Tree roots were exposed like tendons. The forest is spruce and pine, the trail easy to walk on with a layer of needles. The forest is almost soundless, the birds have gone south. Many of the trees show signs of axe work—where a shoulder-high branch has been severed. Mercer cut them last winter.

He's been dividing his time between Trapper Lake and the outside for two decades. In the 1970s he was putting up wooden supports in Cassiar. Then he did the same for a local gold mine. There were many such little operations around Cassiar then and a strong man willing to work could come and go as he pleased. Timbering led to hardrock mining. He worked at Keno. It was a typical mine shantytown, a conglomeration of trailers and lean-tos. Bylaws were as foreign as Hawaiian shirts. Mercer's family was with him at Trapper Lake but he was worried about them getting bushed, so he moved them into a trailer at Keno. The mine manager, an affable guy with a rollercoaster personality, had been to Little Trapper and in Mercer's words "had seen what I was trying to do," so he let him come and go as trapping season dictated.

We rounded a bend and paused by a fallen tree. Hidden in the forest here, the angles so natural and the colours so unremarkable, were the remains of a cabin. Glen Hope and a partner had dog-teamed into the area from Juneau in 1936, looking for new country. Hope was forty. He prospected for six years. His partner left after four.

I was fairly reverential of the site until Mercer toed over a plank, then I started exploring more aggressively. The door of the cabin was shoulder-high. Whether it was made that way or the cabin had sunk was difficult to tell. The sod roof collapsed long ago, as did the root cellar. Remnants of pots and pans were everywhere. The pole bunks were intact. There was a Nestlé baking powder box with a recipe for bread, a finger-jointed butter box, and a saw blade. After passing this site for many years, Mercer took a liking to the knotty tree outside. So he cut it down and started milling it. There were puzzling holes. He poked around and found lead slugs. Glen and his partner must have used it for target practice.

We carried on, following the trail as it veered away from the river. Paralleling and sometimes on the trail were paw prints. Rounded, deeply inset, they were the result of generations of grizzlies passing this way. Grizzlies walk in the steps of other grizzlies; as if they were born to do so, they take the same-sized step with the same foot. I knelt and spread my fingers in the indentation. The prints were the size of a dinner plate.

Following Mercer, I could see how he might cover a lot of ground. Mercer's pace was slow/medium but his strides were long. Every hundred paces or so, I dog-trotted to catch up. He seemed completely oblivious to me, which was both a compliment and deeply worrying. I recalled the story, often told over beers, of a writer who was tarrying behind a hiking group in Alaska. While the others walked ahead, he was harvested by a grizzly. The bear left only a pencil.

After ten minutes we came upon an open area so cleanly delineated that I expected to hear the tinkle of a cowbell or see the peak of a gabled roof. It was about four acres, and rank with bunchgrass, lichen and desiccated mushrooms. Mercer said the meadow grew on alluvial soil and supported voles. He bent and fingered the thatch. It was riven with holes. Voles are food for the pine marten. He said there would be a good harvest of pine marten in two years. When he first came into the country, there were many more open places like this, but the martens were moving. Why this was he did not know.

On the walk back to the skiff Mercer turned and said, "I'm not really in a position to talk. I've worked in mines. But what I worry about is that people who come into this country"—he paused and motioned—"don't have an affinity for the land." We pounded back across the lake in silence. The lake had an in-the-palm feeling. I looked over toward the Thorn. There was the helicopter, like a bird of prey.

The essential tension in any exploration camp is between the budget and exploration. The budget for the Thorn was approximately \$300,000—\$230,000 of which was gone. Combined costs for the camp were about \$10,000 per day. With the budget fast dwindling and the season coming to a close it was crucial that the drillers produce something.

Dave Caulfield had come into the Thorn to help speed up the core logging process. All core has to be logged so the information can be studied and to satisfy security requirements. It is a laborious process.

Caulfield has to duck to go through doorways. He stoops like a sunflower. He's just finished reading a book by a Danish writer that questions many of the arguments environmentalists use to oppose mining. He thinks it is a refreshing counterpoint to the "David Suzukis" who, he feels, oppose any development.

I asked Caulfield how he got into geology. He said that when he got out of high school he didn't know exactly what line of work to get into but he did know that he wanted to work outdoors. He wasn't a fan of biology, so that ruled out forestry. Wandering around the UBC campus one day he went into the geology building and saw a picture of a young woman who had won the student of the year award. "She was beautiful," he said. So he signed on.

Still, his heart wasn't in his studies. His marks were in the high Cs. He took a year off and worked underground drilling. Shifts were twelve hours, sixteen with overtime. "You worked in the dark and came home in the dark. I thought, 'Don't I get to see the sun?'" Once, while underground, he went to wash his boots off in a stream of water shooting out of a drill hole. The driller told him it wasn't a good idea. The next day a drill rod broke off and shot back down the hole. If his boot had been in front of it he would be minus a leg.

Back at UBC for third-year studies, he got the highest marks in the class.

After graduating from UBC he partnered with Henry

Awmack and formed an exploration consulting company. They rented a four-hundred-square-foot office in Vancouver. On the first day of business they walked into the office and realized there was no furniture. A few hours later another exploration company, which was moving to better offices, phoned and said for five hundred dollars they could have desks, drafting table, equipment and two of what Caulfield described as "scuzzy" office chairs. Young and hard-working, Caulfield and Awmack worked long shifts and filed good reports. They bought equipment from older outfits at fifteen cents on the dollar, did their own warehouse work. Soon they hired geologists to help them.

Last night the night-shift drillers, Bill and Lloyd, sunk the drill a further 150 feet. On the morning flight up everyone was optimistic. The helicopter dropped us at a flat area about the size of a tennis court at the junction of La Jaune and Camp Creeks. There was a pile of core here, and two tents. In the bush were the remnants of past exploration camps. With a long line hooked to the belly of his chopper, Norm lifted off and spiralled up and over the creek and hovered over the drill platform. You don't have to be right under a helicopter to have the feeling that, should it drop something, it would fall on your head. Again and again I experienced a kind of reverse vertigo, where I would look up and become fearful.

After a few hectic moments the chopper lifted, the line tightened and the core box lifted off. When the core boxes were on the wooden platform and Norm was gone, we carried them over to a wooden structure that resembled an oversized shoestore footstool. The boxes weighed about one hundred pounds and we took turns carrying them over two by two. The drillers had written an obscene joke on the box and there was nervous laughter. There were eight boxes, each with five rows of five feet long. Caulfield ripped off the lid with a steel claw hammer.

In exploration geology, the drill core represents the ultimate proof of a theory. Core is called the "lie detector." If a theory is sound, the core will prove it—in material evidence. If the theory is incomplete or outright wrong, the core is simply proof that a lot of time and money has been spent for inconclusive results. Caulfield scanned the core quickly. They read it left to right, like a manuscript. There were moments of silence. "It doesn't have quartz veining, doesn't have enargite." He looked longer, then said, "I guess this means we are not millionaires."

For months after I left the Thorn—beside the same silent pilot Moser—I thought about Caulfield and Henry Awmack and Tom Bell and Harry Warren and all the others who looked so hard and earnestly for metal deposits. Why did they do it? A decent answer, I think, is threefold.

First, they do it for the desire for a career that combines middle-class income (geologists rank with senior teachers for pay), opportunity of income (some geologists become very rich) and a job that marries mental and physical challenges with the most striking natural environments in the world.

Second, for the belief, held strongly by many geologists, that they can understand what is going on in rocks better than anyone else. As Henry Awmack wrote to me once, "I think we're all driven by the same motivation: solving the puzzle. We all know that the odds are tremendously against us, but each of us believes that either a) we can put together the bits of information from the stones or the surveys or our knowledge of similar mineralizing systems in a smarter way and understand better than the last guy what's happening in the third dimension (and what was happening in the fourth dimension, 186 million years ago); or b) that we've got sharper eyes, better hunches and more diligent badger abilities than the last guy; or c) that we're in the right spot and that if we just dig a little deeper, or hand-steel that adit another couple of rounds, or . . . we're bound to find it. We all know that there are undiscovered orebodies, hundreds of them, in British Columbia and that *my special talents*, whether in thinking or observing or following hunches or working like a rented mule, *are going to let me succeed where others have failed.*"

But finally, what came into my mind more often than not when I pondered why geologists and prospectors do what they do was the image of a boy sitting cross-legged on the floor of a Manitoba farmhouse, opening a cardboard box containing rocks and fingering each sample.

The boy's name was Cam Stephen and for me the story of Stephen's passion for rocks—embodied and entwined with his life—seems to more satisfactorily answer the questions of why than any rational, measured answer.

By any reasonable standard Cam Stephen is a geologist. Yet he has no degree or doctorate. He knows about rocks and mineral deposits and the people in the industry through a lifetime spent in almost every aspect of the industry. He has worked in mines, exploration, surveying and prospecting.

Stephen was raised on a mixed farm in Petersfield, Manitoba. His father was a former Scottish banker who had lost his job to the Depression. Even as a toddler Stephen showed an interest in rocks, collecting pebbles from a creek bed and scrubbing them clean in the barrel of rainwater reserved for laundry. When Stephen was thirteen his mother took him to the Manitoba museum, triggering in the young man what Stephen calls "every mother's worst nightmare." He thereafter collected rocks, bugs, butterflies—everything. He wrote long letters to the museum about rocks, which they passed on to Professor Edward Leith of the University of Manitoba's geology department. Leith must have recognized something in Stephen, for he sent him rock samples in chocolate boxes and the text *The Outlines of Geology*, and often wrote him encouraging letters.

Pondering his future after high school, Stephen considered working for a Hudson's Bay Company trading post. Then a roast from the local grocery store arrived wrapped in newspaper. Among the ads was one for a surveyor's helper at the Central Patricia Gold Mines, 120 miles north of Sioux Lookout in the Pickle Lake district. He borrowed fifty dollars to get there. "I saw the head frame and felt that I was home," he said.

Still seventeen, he was too young to work underground so they made him timekeeper's assistant. He had to share an adding machine with a man who was right-handed, so Stephen learned to work it left-handed. There were four hundred people in the camp. He transferred to engineering and came under the sway of Mr. and Mrs. Bob Cocaine. Bob Cocaine saw to it that Stephen learned about office organization and survey—all aspects of exploration. Cocaine told Stephen that he would have to learn to drink if he was to succeed—he tutored the shy Stephen in the art by having him run the turntable at camp parties.

The Central Patricia staff included men who had worked during the Yukon Gold Rush and others who would go on to chair geology departments at Canadian universities. He took a ninety-dollar advance on a correspondence course and while the other miners partied he studied. In his spare time he hung out in the crusher plant, learning the processes of milling.

One of his first exploration projects was in Armstrong,

Ontario. A company was interested in a property and wanted Stephen to look at it before winter freeze-up. This was in 1953, and helicopters were being tried for exploration. Stephen and a white-haired, gentlemanly French Canadian met at the airport. The helicopter was a G2, very small. They were used to packing in lots of gear. "The pilot kind of shocked us by saying, 'I can take you one at a time, and you can take a sleeping bag, snowshoes, axe and twenty pounds of groceries.'" They took off in the hangar, rose two feet, then flew out the open hangar doors. The pilot was so worried about the surface of the lake that he had Stephen scramble out of the hovering craft and take an axe to the ice. Only then would he land it.

He left Central Patricia for a job with Combined Developments, an outfit based in Edmonton. Stephen never made it there. The Russians had just exploded a lithium-based bomb and there was a rush to find a source of lithium in North America. Assigned to Manitoba, Stephen walked into the library of the provincial mines department office. On the wall in front of him was a map of Crowduck Bay, scale one mile to the inch. "I have a feeling. I have to go there," he said. He did, and found a lithium deposit. Subsequent drilling confirmed a two-million-ton deposit. It seemed bound for production—until a giant deposit was discovered in Utah two years later.

Stephen was living in Red Deer when he got a call from Dr. Bill Bacon, who was working for Carl Springer on BC properties. Springer, a former trapper in Quebec, was legendary for finding mineral properties. He was also a very hard driver with an encyclopedic knowledge of geology. "It didn't matter where you were, or where you went, he knew it," said Stephens. Springer had asked Bacon to get him the best exploration men. Stephen's name came up again and again. The salary was double what Stephen was getting and the work would be in new, attractive country. On the day Stephen left—it was March 31—he had to shovel snow from seventy-five feet of driveway to get to the road. In Vancouver his future colleagues were playing golf and walking to work in windbreakers.

Though Bacon wanted Stephen to work in the Stikine, Stephen refused. He wasn't competent, he said. Instead, he drove to Williams Lake, where a prospector hoping for a rush had let a huge block of claims lapse. It was one of those decisions that could be called lucky but on closer inspection was more likely to be a result of experience. He met up with what he calls the "right" people: prospectors, road builders, foresters and loggers, each of whom added bits to a picture of the area's geology that Stephen was assembling. One of his colleagues was a former WWII airman who suffered from shellshock. He wasn't able to sleep in hotels or bunkhouses—he preferred a pup tent. Eventually Stephen staked the area that was to become the Bell Mine (later reopened as Mount Polley copper and gold mine).

Working for, and later with, Bacon in various outfits, Stephen became one of the most respected exploration men in BC. Unlike university-trained geologists, whose heads can be full of mineral deposit models, he had heard and seen enough to know that for every theory there was an exception. "You have to have an open mind: what am I really seeing; never mind what someone told me."

Before dinner I walked the ten paces from the cabin to the beach and prospected for skipping stones. The stones here were fist-sized, plum-dark and of all shapes. Few were flat. I like a stone that is the size and shape of a dollar pancake. These were loose and wobbly and made that pleasant clinking sound. I eventually found one that fit in my hand nicely: as thick as my finger, and heavy enough that I won't want to drop it on my toe. I bent to that characteristic stone-skipping posture and let fly. One, two, three bounces then it sloughed to a stone and sunk.

In England, children call the many and repeated bounces that a skipping stone makes "pitty-pats." The game there is known as "ducks and drakes" and is said to loosely imitate the action of waterfowl. The first time the stone skips it is called a duck, the second time a drake, and so on. The game dates to a legendary king who skipped sovereigns across the Thames. Thereafter the phrase came to mean excessive waste, as in "he is ducking and draking away his fortune." To make ducks and drakes with money is to waste it.

George Washington is supposed to have skipped a silver dollar too, but historians have discounted that. Washington, they say, was much too parsimonious to have used money. He probably used a stone.

French physicist Lydéric Bocquet has studied the techniques behind stone skipping—the selection of rock by weight and shape, the throwing posture. His quest was prompted by his son, who asked, What exactly is it that makes for a successful stone skip?

Bocquet was admirably suited to answer the question: he is a physicist with France's Université Claude Bernard Lyon. As it happened, he was looking for examples to use in a mechanics textbook. Bocquet's answer was published in *The American Journal of Physics*. In it, he determined that the chief parameters that determine whether your stone noses under or skims across the lake are the mass of the stone, its angle (lower is better), its spin rate (more is better) and its horizontal velocity.

Armed with calculations of energy loss, Bocquet worked out an expression of the maximum number of skips one can expect.

I picked from the beach another skipping rock—lighter than the last with a worrisome out-of-round outline. The heft in my hand is too light. In the distance the helicopter pounds in. I pull back my arm and let fly. It ploughs in and flashes like a wounded fish on the way to the bottom.

Stones of all sorts—skipping and otherwise—are of such common utility that the idea that some are better than others is absurd. But anyone who has anchored a tarp with a jagged stone will soon find out that a round stone *is* much better. Stones are used as trail markers, as anchor fence posts and road signs, as counterweights in high-tensile fences. To keep the mouth wet and to slake thirst when drink is unavailable, hikers suck on stones. Pumice was once commonly used to redden the cheeks of young women and is still used to rasp calluses. East of the Alleghenies, campers make three-sided walls with rocks and call them ovens; west of the Rockies campers place rocks in a circle. A rock carried in the pocket is considered thumb therapy.

In England, a stone is a unit of measurement: fourteen pounds. When making sauerkraut, the mix needs to be weighted with a stone. The Inuit made soup by boiling lichen-covered stones in a pot. Demosthenes placed stones in his mouth to stop stuttering. With his face to the sky, Pliny watched migrating birds and reported that they must ballast themselves with stones.

At the helicopter clearing I found a cache of skipping stones. Who left this, I wondered? I concocted a small story having to do with a tot and an airplane crash. Sammy waits by the lake collecting skipping stones, but the plane never arrives. He grows up in the valley, a sort of Tarzan of northern BC. He returns to find the stones of his childhood—a reminder of his roots.

It is a B-daydream, granted, but it shares with some enduring stories a similar imagery. Think of the biblical Jacob laying his head down on a stone pillow, David downing Goliath with a stone. Robertson Davies recognized the imagery implicit in stone. That's why he embedded one in a snowball and had a character in the first of his Deptford novels rifle it at an unsuspecting boy. Stones make good beginnings but they make good endings, too. Riven with depression and taunted by voices in her head, Virginia Woolf freighted the pockets of her overcoat with stones and walked into the river.

Stones are the symbol of endurance for many institutions. The Catholic Church is built upon the rock of St. Peter; Peter comes from the French *pierre*, which means "stone." Plymouth Rock is the founding image of the United States of America. In the inner sanctum of the Islamic shrine of Kaaba, in Mecca, is a black stone that may be a meteorite.

From the improbable pile of archeology at my feet I fingered a nicely rounded stone. Granitic, I decided. Or granitoid. The weight was ideal; the arc of its perimeter fit into the crook of my index finger. My thumb folded over the top like a breech. Bending low, I flicked my arm, making sure to concentrate on spinning it. The stone kissed the water one, two, three times, and I knew it was a good shot. It had momentum. It made a dozen skips. Not a world record, but I decided it was a best ever for Little Trapper Lake.

The following movies are about prospecting or gold rushes: *Aguirre, the Wrath of God; Back to the Future Part III; By the Law; The Claim; El Dorado; Eureka; The Gold Rush;*

Mackenna's Gold; North to Alaska; Road to Utopia; The Robin Hood of El Dorado; Support Your Local Sheriff!; A Thousand Pieces of Gold; Way Out West; and *White Fang* (both versions: one by Fulci, the other by Kleiser).

The movie favoured by most critics, and called one of the best films of all time, is John Huston's *Treasure of the Sierra Madre*, in which an old-timer refers to all the gold rushes he's been in on and includes British Columbia. Wrote critic James Agee: It is "one of the most visually alive and beautiful movies I have ever seen; there is a wonderful flow of fresh air, light, vigour and liberty through every shot." The old-timer is played by John Huston's father. It is about three down-on-their-luck prospectors who team up to search for and mine gold, and it includes Humphrey Bogart and a largely forgotten granite-jawed actor named Tim Holt whose father actually was a gold prospector.

I'm not sure why I like the movie so much, though it may have something to do with the fact that the movie has, like gold, many myths attached to it. For example: the author of the book the movie is based on, who is known to history only as B. Traven, was thought to have been a pseudonym of Jack London; or of Ambrose Bierce, who vanished into revolutionary Mexico in 1913, and who was said to be so disfigured by leprosy that he had to hold his head together with a towel. Or, B. Traven was supposedly not an individual but a syndicate of five German writers who lived in the Honduras, including a German-Canadian named Theel or Thiel, or perhaps a Hohenzollern prince.

After a lot of sleuthing, a British film fan tracked down Traven and his story. His real name was Herman Albert Otto Max Wienecke, and one of the few photographs of him show a man of medium build with a wide, downcast mouth and an expression as if he'd forgotten something really, really important. He wrote a number of books and was obsessed with the often-accidental nature of conception and the role of wealth in character disintegration. These themes are expressed clearly in *Treasure of the Sierra Madre*, which is about greed.

Traven was often asked why he wrote *Treasure*, and he often gave different answers. What's clear, however, is that the consequences of wealth were only part of the story—the one that satisfied Hollywood. Beneath the story is a narrative of camaraderie that is nearly Steinbeckian—of working guys eating beans and wearing dirty pants and fighting and getting along. The key to the movie's success is that it addresses the central pleasure in prospecting—the endeavour, not the finding. The movie, like the gold, was Traven's undoing. It fetched on him a kind of attention he had difficulty avoiding.

If I were casting for a remake of *Treasure of the Sierra Madre*, I'd cast Gary Thompson in the role of Tim Holt. Muscular and well-built, with blue eyes and wire-rimmed spectacles, he looks aesthetic, like a fundamentalist preacher. I first heard about Thompson when I passed through Smithers, northbound. A few of the people I talked to said I should meet this crazy guy in Houston who had bought a drill. I had a long list of people to meet, many of whom had more extraordinary recommendations of interest, so I paid little attention. But I kept hearing of Thompson, or this guy in Houston, so on the way south I phoned him. He said he was bound for the bush the next day, but we could meet in Houston at a café called The Elements.

The Elements is the kind of café that people like me (university-educated, sentimental for cafés) will drive a hundred miles for, just to have a piece of unremarkable pie. It is adorned in a way that suggests no aesthetic at all. On the wall were several calendars, which, according to traveller William Least Heat-Moon, are a good sign. The more calendars the better the café. There are several Golden Cup Awards on the wall. The Elements is on the inside of a curve in the highway and the passing of each logging truck sends ripples across the coffee.

I was early; Thompson arrived right on time. He sat down, ordered a coffee, picked up the spoon in his left hand and did not let it go for the next hour. I asked him how he got interested in prospecting. He said he'd driven truck for years, but had an accident one night at a truck stop outside of Alberta. The spring on a mud flap he was adjusting let go and threw him into a ditch. He woke up dazed and confused. His nose was shattered and he almost lost an eye. "That's when I decided to change." He started selling off his equipment, including a Peterbilt truck that he delivered to a customer in Prince George. Listening to the radio on the way home he heard an ad on the radio for a beginner's prospecting course run by Tom Richards.

Thompson said he liked Richards as soon as he saw him. "You could see the knowledge in him and you respected him. When you listen to a guy like that it makes the whole industry huge." The course was held at Northwest Community College in Smithers. It ran for ten days, with some days going up to fifteen hours. One of Thompson's fellow students was a geologist.

Thompson knew he would need money to go prospecting and that no one was likely to pay him without some experience. So he went to work for a local log truck driver, working twenty hours a day. In the months to follow he took more courses, and took field trips to mines: Snip, Golden Bear, Huckleberry, Kemiss and Eskay Creek. Thinking he'd like to know more about drilling, he took a job with Harvey Tremblay. It was hard, dirty work, the type usually done by a farm kid half Thompson's age. But he wanted to know it all. Within a week his hands were so sore he couldn't hold a fork; they sat in his lap like a limp pet. "Everything is fast and hard," Thompson said.

He'd been told, and he believed, that for the first three to four years a prospector needs to see a lot of different ground. So on his days off he'd stop his truck at the side of the road and hike into the bush. In his second year he found an attractive property. "It stayed in the back of my mind—the sizes were there and the indications were there." He let the property lapse but came back. Sitting by the side of the creek, he looked down and saw blue. That was it. The creek was on a fault. There was alteration on both sides of the creek. He registered the property under claim ROX1. Tests returned values of lead, zinc, gold and arsenic. Wanting to find out more, he started digging. He wanted to get to the bedrock. "I thought there should be something there. I had a theory."

He dug an eleven-foot hole. It took two days. He dug another hole into a road shoulder and stopped only when the road threatened to give way. He leaned on his shovel. "I thought: I need an excavator or a drill."

He remembered attending a Cordilleran Roundup conference and seeing a prospector who had core samples, not just soil samples: "He was barely visible from the crowd around him." In the business, this is called something to "take to the table." Thompson was convinced he needed a drill. In August 2001 he drove from Houston to Richmond and purchased a twenty-horsepower Pack Drill, made by Hydracore Drills. Two guys can pack this drill in fifteen loads. For $38,000 he got the drill and all the necessary supplies. None of it rose above the box of his truck. On the way home he stopped at a weigh scale. The entire rig weighed two thousand pounds.

At the end of our interview I wished him luck. Thompson said, "I'm real proud you asked about my story." I said I'd call in a couple of months.

He said, "If I say, 'Well, I'm not driving truck,' you'll know it worked out okay."

The next time I saw him was at that winter's Roundup. He was driving truck, but he had great hopes for the coming exploration season.

Index